Advances in Intelligent Systems and Computing

Volume 1098

AF895991

The series "Advances in Intelligent Systems and Computing" contains publications on theory, applications, and design methods of Intelligent Systems and Intelligent Computing. Virtually all disciplines such as engineering, natural sciences, computer and information science, ICT, economics, business, e-commerce, environment, healthcare, life science are covered. The list of topics spans all the areas of modern intelligent systems and computing such as: computational intelligence, soft computing including neural networks, fuzzy systems, evolutionary computing and the fusion of these paradigms, social intelligence, ambient intelligence, computational neuroscience, artificial life, virtual worlds and society, cognitive science and systems, Perception and Vision, DNA and immune based systems, self-organizing and adaptive systems, e-Learning and teaching, human-centered and human-centric computing, recommender systems, intelligent control, robotics and mechatronics including human-machine teaming, knowledge-based paradigms, learning paradigms, machine ethics, intelligent data analysis, knowledge management, intelligent agents, intelligent decision making and support, intelligent network security, trust management, interactive entertainment, Web intelligence and multimedia.

The publications within "Advances in Intelligent Systems and Computing" are primarily proceedings of important conferences, symposia and congresses. They cover significant recent developments in the field, both of a foundational and applicable character. An important characteristic feature of the series is the short publication time and world-wide distribution. This permits a rapid and broad dissemination of research results.

**** Indexing: The books of this series are submitted to ISI Proceedings, EI-Compendex, DBLP, SCOPUS, Google Scholar and Springerlink ****

More information about this series at http://www.springer.com/series/11156

M. Arif Wani · Mehmed Kantardzic ·
Moamar Sayed-Mouchaweh
Editors

Deep Learning Applications

Springer

Editors
M. Arif Wani
Post Graduate Department
of Computer Science
University of Kashmir
Srinagar, Jammu and Kashmir, India

Mehmed Kantardzic
Department of Computer Engineering
and Computer Science
University of Louisville
Louisville, USA

Moamar Sayed-Mouchaweh
Department of Computer Science
and Automatic Control
High National Engineering School
of Mines Telecom Lille Douai
Douai, France

ISSN 2194-5357 ISSN 2194-5365 (electronic)
Advances in Intelligent Systems and Computing
ISBN 978-981-15-1815-7 ISBN 978-981-15-1816-4 (eBook)
https://doi.org/10.1007/978-981-15-1816-4

This Springer imprint is published by the registered company Springer Nature Singapore Pte Ltd.
The registered company address is: 152 Beach Road, #21-01/04 Gateway East, Singapore 189721, Singapore

Preface

Machine learning technology is a new digital frontier that may transform many facets of a modern society. It is expected to have a profound impact on the world, transforming the way we live, work, shop, travel and interact. Machine learning-based inventions are booming, shifting from theory to commercial application, supporting important developments in science, technology and business, from autonomous vehicles and medical diagnosis to advanced manufacturing and Internet-based recommender systems. Deep learning techniques represent the frontiers in the machine learning innovations with tremendous success in a variety of applications. Unlike conventional machine learning techniques, deep learning is able to generate automatically high-level data representations from massive volume of raw data. Therefore, it has provided a solution to many real-world applications, particularly with the increased processing power and the advances in graphics processors.

Recent breakthroughs in deep learning applied to speech recognition, computer vision and machine translation have been so outstanding, with algorithms functioning close to human performances, and enabling clear impacts on industry, society and economy. Deep learning technology includes a large number of different deep architectures, from deep feedforward networks (DFNNs) and restricted Boltzmann machines (RBMs), through deep belief networks (DBNs) and auto-encoders (AE), to convolutional neural networks (CNNs), recurrent neural networks (RNNs) and generative adversarial networks (GANs). The Internet, financial institutions and e-commerce are the biggest industries that are already affected by this new technology, but we will soon recognize new advancements in the application areas such as retail, health care, manufacturing, transportation, agriculture and logistics. The combination of deep learning technologies with other emerging technologies, such as robotics, Internet of Things and cryptography used for blockchain, may revolutionize other areas and lead to further expansions in new domains. Deep learning applications could help trigger structural transformation in local economies, such as the modernization of traditional sectors, the diversification of industries facing declining markets or the transition of a sector toward more productive activities. Soon, we will not be able to recognize an industry that is not being touched by deep learning; many users do not realize that they currently interact with deep learning products almost every day.

Since deep learning-related innovations are enabled by data, the organizations that generate the most advanced products are often the ones that own the most data. That is the main reason why the technology has attracted the attention of many high-tech enterprises such as Google, Facebook and Microsoft. But, today there are hundreds of small companies and many successful start-ups which represent promising new support for the fast developing field.

Deep learning technology is complex and potentially affects many different areas of human activities, especially because it is related to the large amount of multimedia data. Concerns around data used for deep learning are becoming more central, and include ethical questions, from the fear of security breaches and hacking, issues around privacy, trust and autonomy of these new systems, to potential bias in deep learning algorithms. When deep learning-supported medical diagnostic platforms are trained on poorly curated data, it can result in unsafe diagnostics or treatment interventions that no doctor would ever recommend for use. Therefore, deep learning requires to make data quality standards and model auditing procedures of the utmost importance. Current deep learning applications will also need to adapt to the rising issues such as data sparsity, missing data, incomplete, unlabeled, heterogeneous and messy data.

We are a part of the first major wave in a deep learning revolution, and soon, we will see more and more impacts of this technology on our lives. The main intention of this book is to explore and present some new applications of deep learning covering a variety of areas, techniques and deep learning architectures used. Illustrative examples, which are presented in the book, are small but additional contribution to the field in understanding the importance of real-world deep learning applications.

Srinagar, India
Louisville, USA
Douai, France

M. Arif Wani
Mehmed Kantardzic
Moamar Sayed-Mouchaweh

Contents

Trends in Deep Learning Applications 1
M. Arif Wani, Mehmed Kantardzic and Moamar Sayed-Mouchaweh

Quasi-Newton Optimization Methods for Deep Learning Applications 9
Jacob Rafati and Roummel F. Marica

Medical Image Segmentation Using Deep Neural Networks with Pre-trained Encoders 39
Alexandr A. Kalinin, Vladimir I. Iglovikov, Alexander Rakhlin and Alexey A. Shvets

Ensemble of 3D Densely Connected Convolutional Network for Diagnosis of Mild Cognitive Impairment and Alzheimer's Disease 53
Shuqiang Wang, Hongfei Wang, Albert C. Cheung, Yanyan Shen and Min Gan

Detecting Work Zones in SHRP2 NDS Videos Using Deep Learning Based Computer Vision 75
Franklin Abodo, Robert Rittmuller, Brian Sumner and Andrew Berthaume

Action Recognition in Videos Using Multi-stream Convolutional Neural Networks 95
Helena de Almeida Maia, Darwin Ttito Concha, Helio Pedrini, Hemerson Tacon, André de Souza Brito, Hugo de Lima Chaves, Marcelo Bernardes Vieira and Saulo Moraes Villela

Deep Active Learning for Image Regression 113
Hiranmayi Ranganathan, Hemanth Venkateswara, Shayok Chakraborty and Sethuraman Panchanathan

Deep Learning Based Hazard Label Object Detection for Lithium-ion Batteries Using Synthetic and Real Data 137
Anneliese Schweigert, Christian Blesing and Christoph M. Friedrich

Enabling Robust and Autonomous Materialhandling in Logistics Through Applied Deep Learning Algorithms 155
Christian Poss, Thomas Irrenhauser, Marco Prueglmeier, Daniel Goehring, Vahid Salehi and Firas Zoghlami

Author Index 177

About the Editors

Prof. M. Arif Wani completed his M.Tech. in Computer Technology at the Indian Institute of Technology, Delhi, and his Ph.D. in Computer Vision at Cardiff University, UK. Currently, he is a Professor at the University of Kashmir, having previously served as a Professor at California State University Bakersfield. His main research interests are in gene expression datasets, face recognition techniques/ algorithms, artificial neural networks, and deep architectures. He has published many papers in reputed journals and conferences in these areas. He was honored with the International Technology Institute Award in 2002 by the International Technology Institute, California, USA. He is a member of many academic and professional bodies, e.g., the Indian Society for Technical Education, Computer Society of India, IEEE USA, and Optical Society of America.

Dr. Mehmed Kantardzic received his Ph.D. in Computer Science in 1980, M.S. in Computer Science in 1976, and B.S. in Electrical Engineering in 1972, all from the University of Sarajevo, Bosnia, and Herzegovina. He served as an Assistant, and Associate Professor at the University of Sarajevo, and later as Associate and since 2004 as Full Professor at the University of Louisville. Currently, he is the Director of the Data Mining Lab as well as the Director of CECS Graduate Studies at the CECS Department. His research focuses on data mining & knowledge discovery, machine learning, soft computing, click fraud detection and prevention, concept drift in streaming data, and distributed intelligent systems. Dr. Kantardzic is the author of six books including the textbook: "Data Mining: Concepts, Models, Methods, and Algorithms" (John Wiley, second edition, 2011) which is accepted for data mining courses at more than hundred universities in USA and abroad. He is the author of over 40 peer-reviewed journal publications, 20 book chapters, and over 200 reviewed articles in the proceedings of international conferences. His recent research projects are supported by NSF, KSTC, US Treasury Department, U.S. Army, and NASA. Dr. Kantardzic was selected as

Fulbright Specialist in Information Technology in 2012. Dr. Kantardzic has served on the Editorial Boards for several international journals, and he is currently an Associate Editor for WIREs Data Mining and Knowledge Discovery Journal.

Prof. Moamar Sayed-Mouchaweh received his Ph.D. from the University of Reims, France. He was working as Associated Professor in Computer Science, Control, and Signal Processing at the University of Reims, France, in the Research Centre in Sciences and Technology of the Information and the Communication. In December 2008, he obtained the Habilitation to Direct Research (HDR) in Computer science, Control, and Signal Processing. Since September 2011, he is working as a Full Professor in the High National Engineering School of Mines Telecom Lille Douai (France), Department of Computer Science and Automatic Control. He edited and wrote several Springer books and served as a Guest Editor of several special issues of international journals. He also served as IPC Chair and Conference Chair of several international workshops and conferences. He is serving as a member of the Editorial Board of several international journals.

Trends in Deep Learning Applications

M. Arif Wani, Mehmed Kantardzic and Moamar Sayed-Mouchaweh

Abstract Deep Learning is a new area of Machine Learning which has gained popularity in the recent past. It has surpassed the conventional algorithms in accuracy as the features are learned from the data using a general purpose learning procedure instead of being designed by human engineers [1]. Deep learning is responsible for today's explosion of AI. Deep networks have demonstrated dramatic improvements in computer vision and machine translation tasks. It has the ability to recognize spoken words nearly as good as humans can. It has demonstrated good generalization power and has achieved high accuracy in machine learning modeling, which has even attracted non-computer scientists. It is now being used as a guide to make key decisions in fields like medicine, finance, manufacturing, and beyond. Deep learning has succeeded in previously unsolved problems which were quite difficult to resolve using machine learning as well as other shallow networks. However, deep learning is still in its infancy, but it is likely that deep learning will have many successes in the near future as it requires little hand engineering and thus can take advantage of the vast amount of data and computation power. Various supervised and unsupervised deep architectures have been reported in [1]. This chapter outlines the use of deep learning technology in applications like game playing, medical applications, video analytics, regression/classification, object detection/recognition, and robotic automation.

M. Arif Wani (✉)
University of Kashmir, Hazratbal, Srinagar 190006, India
e-mail: awani@uok.edu.in

M. Kantardzic
University of Louisville, Louisville, KY, USA
e-mail: mmkant01@louisville.edu

M. Sayed-Mouchaweh
High National Engineering School of Mines Telecom Lille Douai, Douai Cedex, France
e-mail: moamar.sayed-mouchaweh@mines-douai.fr

M. A. Wani et al. (eds.), *Deep Learning Applications*,
Advances in Intelligent Systems and Computing 1098,
https://doi.org/10.1007/978-981-15-1816-4_1

1 Introduction

Deep Learning is a new area of Machine Learning which has gained popularity in the recent past. It has surpassed the conventional algorithms in accuracy as the features are learned from the data using a general purpose learning procedure instead of being designed by human engineers [1]. Deep learning is responsible for today's explosion of AI. Deep networks have demonstrated dramatic improvements in computer vision and machine translation tasks. It has the ability to recognize spoken words nearly as good as humans can. It has demonstrated good generalization power and has achieved high accuracy in machine learning modeling, which has even attracted non-computer scientists. It is now being used as a guide to make key decisions in fields like medicine, finance, manufacturing, and beyond.

Deep learning has succeeded in previously unsolved problems which were quite difficult to resolve using machine learning as well as other shallow networks. However, deep learning is still in its infancy, but it is likely that deep learning will have many successes in the near future as it requires little hand engineering and thus can take advantage of the vast amount of data and computation power.

Various supervised and unsupervised deep architectures have been reported in [1]. This chapter outlines the use of deep learning technology in applications like game playing, medical applications, video analytics, regression/classification, object detection/recognition, and robotic automation.

2 Deep Learning in Game Playing

The first-order Stochastic gradient descent (SGD) algorithms used for solving optimization problems in deep learning may have slow convergence rates. Finding search directions by using second-order curvature information can help with more robust convergences. However, it requires computing Hessian matrices which are not computationally practical. Quasi-Newton methods can construct an approximation of the Hessian matrix. The quasi-Newton methods, like SGD, require only first-order gradient information, and it can result in good linear convergence. The limited-memory Broyden–Fletcher–Goldfarb–Shanno (L-BFGS) approach is one of the most popular quasi-Newton methods that construct positive definite Hessian approximations, which can be applied to optimization problems in deep learning. The method can be evaluated for various applications including game playing.

Chapter 2 presents efficient optimization methods that are based on L-BFGS quasi-Newton methods as an alternative to the SGD methods used to train deep neural networks. The methods use trust-region strategies to compute low-rank updates to Hessian approximations. The chapter compares the line search L-BFGS optimization method with the proposed Trust-Region Minimization Algorithm. The two methods are implemented to train the LeNet-5 architecture. MNIST dataset has been used to compare the performance of two methods. The chapter presents an optimization method based on line search limited-memory BFGS for the deep reinforcement learning framework. The method is evaluated on six classic ATARI 2600 games. The results show a robust convergence with preferred generalization characteristics as well as fast training time.

3 Deep Learning in Medical Applications

Segmentation of images using deep neural networks has recently received a great deal of attention. Deep Learning-based segmentation has been applied to many applications which include medical imaging and producing improved performance results. Segmentation of medical images is an important machine vision task that often has to be performed to enable computer-aided diagnostics and other medical analyses. However, the segmentation of medical images still remains a challenging problem, but the application of deep learning can result in performance improvements that can potentially benefit the diagnosis and other clinical practice outcomes.

Chapter 3 presents applications of multiple deep convolutional neural networks to medical image segmentation. First, angiodysplasia lesion segmentation from wireless capsule endoscopy videos is presented. Angiodysplasia is the most common vascular lesion of the gastrointestinal tract in the general population. For this application, the dataset consists of 1200 color images obtained with wireless capsule endoscopy. The images are in 24-bit PNG format, with 576 × 576 pixel resolution. In the second application, a model is developed for the semantic segmentation of robotic instruments in surgical videos, challenging problem very important for intra-operative guidance. The dataset used for modeling consists of 8 × 225-frame sequences of high-resolution stereo camera images acquired from a da Vinci Xi. Every video sequence consists of two stereo channels taken from left and right cameras and has a 1920 × 1080 pixel resolution in RGB format. Four different deep architectures for segmentation are analyzed: U-Net, two modifications of TernausNet, and LinkNet-34. In all cases, the output of the model is an image, in which each pixel value corresponds to a probability of belonging to the area of interest or a class. LinkNet-34 was the best performing model for angiodysplasia lesion segmentation that also provides the fastest inference speed, while the best results for semantic segmentation of robotic instruments in surgical videos were obtained by TernausNet-16 model with pre-trained VGG-16 encoder.

Alzheimer's disease is a common progressive neurodegenerative disease, which has been listed as the fourth biggest killer threatening the life and health of the elderly, while mild cognitive impairment has been known as an intermediate transition between normal elderly and person with Alzheimer's disease. Automatic diagnosis of Alzheimer's disease and mild cognitive impairment from 3D brain magnetic resonance images plays an important approach in the early treatment of dementia disease.

Chapter 4 presents deep learning methodology which is based on an ensemble of 3D densely connected convolutional networks for dementia diagnosis from 3D

MRIs. Dense connections were introduced to maximize the information flow, and the probability-based ensemble approach decreases the misrecognition risk of selecting a single classifier. The study involves more than 1,000 participants including people with mild cognitive impairment, patients diagnosed with Alzheimer's disease, and people without the disease. Most of the participants were collected repeatedly for two to six times, and the interval between consecutive scans was more than a year. Multiple experiments were conducted to optimize the performance of 3D-DenseNet with different hyper-parameters and architectures. The results show that, compared with the traditional method based on 2D image slices, 3D convolution retains and extracts more key spatial feature information from brain MRI images, thus providing better features basis for Alzheimer's disease classification model.

4 Deep Learning in Video Analytics

The second Strategic Highway Research Program (SHRP2) funds a number of transportation safety-related projects including the Naturalistic Driving Study (NDS). The study has collected driver- and vehicle-specific data from human drivers in a wide variety of real-world roadway and environmental conditions. Volunteer driver vehicles were outfitted with two external and two internal cameras, forward facing radar, and a Data Acquisition System enabling the development of machine learning models for pedestrian detection, scene segmentation, traffic signal state detection, head, torso, and hand pose detection, and facial feature detection. A complementary project to the NDS was the creation of the Roadway Information Database (RID), which contains stationary characteristics of the portions of the U.S. roadway network over which the NDS drivers drove, such as the number of lanes, the type of turn lane at an intersection, the presence or absence of rumble strips on a highway segment, and so on. Since the inception of SHRP2, the presence or absence of work zones in a given NDS trip has been a desired piece of information.

Chapter 5 presents the new video analytics application, which aims to perform a complete and accurate accounting of work zone occurrences in the video data set collected through NDS. Most data were collected between 2012 and 2015 by more than 3,500 drivers across six states, spanning over five million trips and resulting in more than one million hours of recorded video. A total of 1344 videos containing 31,535,862 frames were made available for use as sources for data set construction, including training, validation, and test subsets. The proposed methodology combines a video decoder based on FFmpeg, an image scene classifier based on TensorFlow, and algorithms for reading timestamps off of video frames, for identifying the start and end timestamps of detected work zone events. Using an active learning approach, a model is initially trained on a small hand-labeled training set of "seed" samples, then use to predict the classes of the remaining unlabeled samples. The active learning approach is particularly attractive, even critical for the success of the application because work zone features are relatively infrequent in a database causing time wasted labeling. The implemented deep learning solution adds to the Roadway Information

Database (RID) the ability to formulate a variety of questions and analyze a specific situation related to construction zones.

The recognition of human actions in videos that aim to detect and identify actions of one or more agents can be a challenging task with difficult scene conditions, such as occlusion, background clutter, and camera motion, and when the same action may vary with a different actor.

Chapter 6 proposes a deep convolutional neural networks (CNNs) in order to recognize human actions in videos. In order to address the challenges related to this recognition, such as occlusions, background clutter, and camera motion, the proposed approach uses three complementarity modalities: RGB frames, optical flow images, and visual rhythm. The first two modalities allow exploring, respectively, the spatial and temporal properties of objects in the consecutive frames while the third one allows to encode the entire video (spatio-temporal) in a single image. The proposed approach is based on the use of three deep CNNs, one for each modality. Each CNNs is fine-tuned with its corresponding modality. Three different fusion strategies (simple average, weighted average, external fully connected layers) are applied in order to obtain a single score vector for the input video. The use of external fully connected layers allows defining how much the features (modalities) contribute to the final prediction. The proposed approach was tested using video sequences collected from commercial and non-commercial sources including blurred videos, or with lower quality and actions from different points of view. The obtained results are promising compared to the state-of-the-art approaches.

5 Deep Learning in Regression and Classification

Training a CNN model for classification/regression task requires large amounts of labeled training data, which is time-consuming and expensive to acquire. Active learning algorithms automatically identify a subset of representative instances from large amounts of unlabelled data, which reduces human annotation effort. This also reduces computational effort to train a machine learning model.

Chapter 7 integrates an active learning mechanism and deep convolutional neural networks for image regression task. The goal of this integration is to perform regression with a small amount of labeled data and a large amount of unlabeled data. Indeed, the active learning mechanism allows selecting salient and informative samples within a huge number of unlabeled data in order to be labeled manually by oracles. This reduces significantly the labeling costs (human efforts) while improving the generalization capacity of the regression model. The criterion used to select the informative or exemplar samples in the proposed mechanism is called the principle of expected model output change (EMOC). The latter quantifies the importance of a sample by measuring the difference between the regression model output when trained with and without a particular sample. This operation is repeated until reaching a predefined number of iterations. The error-based performances (Mea Squared Error, Mean Absolute Error) of the proposed active learning CNN approach is compared with some well-known state-of-the-art regression-based active learning algorithms (Greedy, Query-by-Committee, Random Sampling, Käding) using different data sets (hand-written digit recognition, head pose, age recognition). The obtained results showed that the proposed approach outperforms the other compared state-of-the-art approaches.

6 Deep Learning in Object Detection and Recognition

One of the applications of deep learning can be in hazard label object detection. Different modes of carriers like air, road, water, and rail are used to transport goods which may include dangerous or hazardous items, such as lithium-ion batteries. Inadequate care and inappropriate handling of parcels with dangerous or hazardous items may cause accidents or lead to explosions and fires. In order to separate such parcels automatically, labels of such parcels have to be recognized. Label recognition may become difficult if the hazard labels may get smug by environmental influences during transport.

Chapter 8 presents Convolutional Neural Networks (CNN) based systems to address this challenge is investigated. The systems are based on You Only Look Once (YOYO) and self-developed Object Detection Pipeline (ODP) using Maximally Stable Extremal Regions (MSER). Different evaluation criteria, such as mean average precision (mAP), detection speed, and training time, are used to investigate the performances of these systems for symbol detection of hazard labels on parcels. Two models are trained using datasets of real images and synthetic images that are generated with random contrast, brightness, and blur adjustments.

7 Deep Learning in Robotic Automation

Logistics activities cause almost 25% of the total cost of manufacturer purchase prices. The remaining 75% is mainly composed of material costs (60%) and production costs (15%). The high-dynamic production processes in the modern industry increase the complexity of logistics and thus encourages the growing adoption of automation solutions. Therefore, robotic solutions and intelligent machines are transforming manufacturing industries to a higher level of automation.

Chapter 9 investigates the use of a modular intelligent perception algorithm in order to automate the entire intralogistics using robots. Intralogistics comprises the organization, control, implementation, and optimization of internal material flows, information flows, and goods handling in industry, trade, and public institutions. The proposed algorithm is based on three connected modules the detection, selection, and localization. The detection module aims at identifying the objects of the searched classes in the field of vision of the robot. The output of this module is the set of objects detected in the corresponding image of the searched class. The second module, selection module, selects the relevant objects from this set for the respective process. The output of this second module is the set of objects to be gripped by the robot. Finally, the third module, localization, determines the exact gripping pose in three dimensional space. The performance (successful grips rate) of the proposed perception algorithm is tested using a depalletizing robot in a real process environment.

Reference

1. M. Arif Wani, F.A. Bhat, S. Afzal, A. Khan, *Advances in Deep Learning*, vol. 57 (Springer, Berlin, 2020)

Quasi-Newton Optimization Methods for Deep Learning Applications

Jacob Rafati and Roummel F. Marica

Abstract Deep learning algorithms often require solving a highly nonlinear and non-convex unconstrained optimization problem. Methods for solving optimization problems in large-scale machine learning, such as deep learning and deep reinforcement learning (RL), are generally restricted to the class of first-order algorithms, like stochastic gradient descent (SGD). While SGD iterates are inexpensive to compute, they have slow theoretical convergence rates. Furthermore, they require exhaustive trial-and-error to fine-tune many learning parameters. Using second-order curvature information to find search directions can help with more robust convergence for non-convex optimization problems. However, computing Hessian matrices for large-scale problems is not computationally practical. Alternatively, quasi-Newton methods construct an approximate of the Hessian matrix to build a quadratic model of the objective function. Quasi-Newton methods, like SGD, require only first-order gradient information, but they can result in superlinear convergence, which makes them attractive alternatives to SGD. The limited-memory Broyden–Fletcher–Goldfarb–Shanno (L-BFGS) approach is one of the most popular quasi-Newton methods that constructs positive definite Hessian approximations. In this chapter, we propose efficient optimization methods based on L-BFGS quasi-Newton methods using line search and trust-region strategies. Our methods bridge the disparity between first- and second-order methods by using gradient information to calculate low-rank updates to Hessian approximations. We provide formal convergence analysis of these methods as well as empirical results on deep learning applications, such as image classification tasks and deep reinforcement learning on a set of Atari 2600 video games. Our results show a robust convergence with preferred generalization characteristics as well as fast training time.

J. Rafati (✉) · R. F. Marica
University of California, 5200 North Lake Road, Merced, CA 95343, USA
e-mail: jrafatiheravi@ucmerced.edu
URL: http://rafati.net

R. F. Marica
e-mail: rmarcia@ucmerced.edu
URL: http://faculty.ucmerced.edu/rmarcia

M. A. Wani et al. (eds.), *Deep Learning Applications*,
Advances in Intelligent Systems and Computing 1098,
https://doi.org/10.1007/978-981-15-1816-4_2

1 Introduction

Deep learning (DL) is becoming the leading technique for solving large-scale machine learning (ML) problems, including image classification, natural language processing, and large-scale regression tasks [1, 2]. Deep learning algorithms attempt to train a function approximation (*model*), usually a deep convolutional neural network (CNN), over a large dataset. In most of the deep learning and deep reinforcement learning (RL) algorithms, solving an *empirical risk minimization* (ERM) problem is required [3]. The ERM problem is a highly nonlinear and non-convex unconstrained optimization problem of the form

$$\min_{w\in\mathbb{R}^n} \mathcal{L}(w) \triangleq \frac{1}{N}\sum_{i=1}^{N} \ell_i(w), \tag{1}$$

where $w \in \mathbb{R}^n$ is the vector of trainable parameters of the CNN model, n is the number of the learning parameters, N is the number of observations in a training dataset, and $\ell_i(w) \triangleq \ell(w; x_i, y_i)$ is the error of the current model's prediction for the ith observation of the *training* dataset, $\mathcal{D} = \{(x_i, y_i) \mid i = 1, \ldots, N\}$.

1.1 Existing Methods

Finding an efficient optimization algorithm for the large-scale, non-convex ERM problem (1) has attracted many researchers [1]. There are various algorithms proposed in the machine learning and optimization literatures to solve (1). Among those, one can name first-order methods such as stochastic gradient descent (SGD) methods [4–7], the quasi-Newton methods [8–11], and also Hessian-free methods [12–15].

Since, in large-scale machine learning problems usually, N and n are very large numbers, the computation of the true gradient $\nabla\mathcal{L}(w)$ is expensive and the computation of the true Hessian $\nabla^2\mathcal{L}(w)$ is not practical. Hence, most of the optimization algorithms in machine learning and deep learning literature are restricted to variants of first-order gradient descent methods, such as SGD methods. SGD methods use a small random sample of data, $J_k \in \mathcal{S}$, to compute an approximate of the gradient of the objective function, $\nabla\mathcal{L}^{(J_k)}(w) \approx \nabla\mathcal{L}(w)$. At each iteration of the learning update, the parameters are updated as $w_{k+1} \leftarrow w_k - \eta_k\nabla\mathcal{L}^{(J_k)}(w_k)$, where η_k is referred to as the *learning rate*.

The computational cost-per-iteration of SGD algorithms is small, making them the most widely used optimization methods for the vast majority of deep learning applications. However, these methods require fine-tuning of many hyperparameters, including the learning rates. The learning rate is usually chosen to be very small; therefore, the SGD algorithms require revisiting many epochs of data during the learning process. Indeed, it is unlikely that the SGD methods perform successfully in the first attempt at a problem, though there is recent work that addresses tuning hyperparameters automatically (see e.g., [16, 17]).

Another major drawback of SGD methods is that they struggle with saddle-points that occur in most non-convex optimization problems. These saddle-points have an undesirable effect on the model's generalization ability. On the other hand, using the second-order curvature information can help produce more robust convergence. Newton's method, which is a second-order method, uses the Hessian, $\nabla^2\mathcal{L}(w)$, and the gradient to find the search direction, $p_k = -\nabla^2\mathcal{L}(w_k)^{-1}\nabla\mathcal{L}(w_k)$. A line search strategy is often used to find the step length along the search direction to guarantee convergence. The main bottleneck in second-order methods is the serious computational challenges involved in the computation of the Hessian, $\nabla^2\mathcal{L}(w)$, for large-scale problems, in which it is not practical because n is large. Quasi-Newton methods and Hessian-free methods both use approaches to approximate the Hessian matrix without computing and storing the true Hessian matrix. Specifically, Hessian-free methods attempt to find an approximate Newton direction by solving $\nabla^2\mathcal{L}(w_k)p_k = -\nabla\mathcal{L}(w_k)$ without forming the Hessian using conjugate-gradient methods [12–15].

Quasi-Newton methods form an alternative class of first-order methods for solving the large-scale non-convex optimization problem in deep learning. These methods, as in SGD, require only computing the first-order gradient of the objective function. By measuring and storing the difference between consecutive gradients, quasi-Newton methods construct *quasi-Newton matrices* $\{B_k\}$ which are low-rank updates to the previous Hessian approximations for estimating $\nabla^2\mathcal{L}(w_k)$ at each iteration. They build a quadratic model of the objective function by using these quasi-Newton matrices and use that model to find a sequence of search directions that can result in superlinear convergence. Since these methods do not require the second-order derivatives, they are more efficient than Newton's method for large-scale optimization problems [18].

There are various quasi-Newton methods proposed in the literature. They differ in how they define and construct the quasi-Newton matrices $\{B_k\}$, how the search directions are computed, and how the parameters of the model are updated [9, 18–20].

1.2 Motivation

The Broyden–Fletcher–Goldfarb–Shanno (BFGS) method [21–24] is considered the most widely used quasi-Newton algorithm, which produces a positive definite matrix B_k for each iteration. The conventional BFGS minimization employs *line search*, which first attempts to find the search directions by computing $p_k = -B_k^{-1}\nabla\mathcal{L}(w_k)$ and then decides on the step size $\alpha_k \in (0, 1]$ based on sufficient decrease and curvature conditions [18] for each iteration k and then update the parameters $w_{k+1} = w_k + \alpha_k p_k$. The line search algorithm first tries the unit step length $\alpha_k = 1$ and if it does not satisfy the sufficient decrease and the curvature conditions, it recursively reduces α_k until some stopping criteria (for example $\alpha_k < 0.1$).

Solving $B_k p_k = -\nabla\mathcal{L}(w_k)$ can become computationally expensive when B_k becomes a high-rank update. The *limited-memory* BFGS (L-BFGS) method constructs a sequence of low-rank updates to the Hessian approximation; consequently solving $p_k = B_k^{-1}\nabla\mathcal{L}(w_k)$ can be done efficiently. As an alternative to gradient descent, limited-memory quasi-Newton algorithms with line search have been implemented in a deep learning setting [25]. These methods approximate second derivative information, improving the quality of each training iteration and circumventing the need for application-specific parameter tuning.

There are computational costs associated with the satisfaction of the sufficient decrease and curvature conditions, as well as finding α_k using line search methods. Also if the curvature condition is not satisfied for $\alpha_k \in (0, 1]$, the L-BFGS matrix may not stay positive definite, and the update will become unstable. On the other hand, if the search direction is rejected in order to preserve the positive definiteness of L-BFGS matrices, the progress of learning might stop or become very slow.

Trust-region methods attempt to find the search direction, p_k, in a region within which they trust the accuracy of the quadratic model of the objective function, $\mathcal{Q}(p_k) \triangleq \frac{1}{2}p_k^T B_k p_k + \nabla\mathcal{L}(w_k)^T p_k$. These methods not only have the benefit of being independent of the fine-tuning of hyperaparameters but may also improve upon the training performance and the convergence robustness of the line search methods. Furthermore, trust-region L-BFGS methods can easily reject the search directions if the curvature condition is not satisfied in order to preserve the positive definiteness of the L-BFGS matrices [26]. The computational bottleneck of trust-region methods is the solution of the trust-region subproblem. However, recent work has shown that the trust-region subproblem can be efficiently solved if the Hessian approximation, B_k, is chosen to be a quasi-Newton matrix [20, 27].

1.3 Applications and Objectives

Deep learning algorithms attempt to solve large-scale machine learning problems by learning a model (or a parameterized function approximator) from the observed data in order to predict the unseen events. The model used in deep learning is an artificial neural network which is a stack of many convolutional layers, fully connected layers, nonlinear activation functions, etc. Many data-driven or goal-driven applications can be approached by deep learning methods. Depending on the application and the data, one should choose a proper architecture for the model and define an empirical loss function. (For the state-of-the-art deep neural network architectures, and deep learning applications for supervised learning and unsupervised learning, see [2].) The common block in all deep learning algorithms is the optimization step for solving the ERM problem defined in (1).

In this chapter, we present methods based on quasi-Newton optimization for solving the ERM problem for deep learning applications. For numerical experiments, we focus on two deep learning applications, one in supervised learning and the other one in reinforcement learning. The proposed methods are general purpose and can be employed for solving optimization steps of other deep learning applications.

First, we introduce novel large-scale L-BFGS optimization methods using the trust-region strategy—as an alternative to the gradient descent methods. This method is called Trust-Region Minimization Algorithm for Training Responses (TRMinATR) [26]. We implement practical computational algorithms to solve the Empirical Risk Minimization (ERM) problems that arises in machine learning and deep learning applications. We provide empirical results on the classification task of the MNIST dataset and show robust convergence with preferred generalization characteristics. Based on the empirical results, we provide a comparison between the trust-region strategy with the line search strategy on their different convergence properties. TRMinATR solves the associated trust-region subproblem, which can be computationally intensive in large-scale problems, by efficiently computing a closed form solution at each iteration. Based on the distinguishing characteristics of trust-region algorithms, unlike line search methods, the progress of the learning will not stop or slow down due to the occasional rejection of the undesired search directions. We also study techniques for initialization of the positive definite L-BFGS quasi-Newton matrices in the trust-region strategy so that they do not introduce any false curvature conditions when constructing the quadratic model of the objective function.

Next, we investigate the utility of quasi-Newton optimization methods in deep reinforcement learning (RL) applications. RL—a class of machine learning problems —is learning how to map situations to actions so as to maximize numerical reward signals received during the experiences that an artificial agent has as it interacts with its environment [28]. An RL agent must be able to sense the state of its environment and must be able to take actions that affect the state. The agent may also be seen as having a goal (or goals) related to the state of the environment. One of the challenges that arise in real-world reinforcement learning (RL) problems is the "curse of dimensionality". Nonlinear function approximators coupled with RL have made it possible to learn abstractions over high-dimensional state spaces [29–34]. Successful examples of using neural networks for RL include learning how to play the game of Backgammon at the Grand Master level [35]. More recently, researchers at DeepMind Technologies used a deep Q-learning algorithm to play various Atari games from the raw screen image stream [36, 37]. The deep Q-learning algorithm [36] employed a convolutional neural network (CNN) as the state-action value function approximation. The resulting performance on these games was frequently at or better than the human level. In another effort, DeepMind used deep CNNs and a Monte Carlo Tree Search algorithm that combines supervised learning and RL to learn how to play the game of Go at a superhuman level [38].

We implement an L-BFGS optimization method for deep reinforcement learning framework. Our deep L-BFGS Q-learning method is designed to be efficient for parallel computation using GPUs. We investigate our algorithm using a subset of the Atari 2600 games, assessing its ability to learn robust representations of the state-action value function, as well as its computation and memory efficiency. We also analyze the convergence properties of Q-learning using a deep neural network employing L-BFGS optimization.

1.4 Chapter Outline

In Sect. 2, we introduce a brief background on machine learning, deep learning, and optimality conditions for unconstrained optimization. In Sect. 3, we introduce two common optimization strategies for unconstrained optimization, i.e., line search and trust-region. In Sect. 4, we introduce quasi-Newton methods based on L-BFGS optimization in both line search and trust-region strategies. In Sect. 5, we implement algorithms based on trust-region and line search L-BFGS for the image recognition task. In Sect. 6 we introduce the RL problem and methods based on L-BFGS line search optimization for solving the ERM problem in deep RL applications.

2 Unconstrained Optimization Problem

In the unconstrained optimization problem, we want to solve the minimization problem (1).

$$\min_{w} \mathcal{L}(w), \tag{2}$$

where $\mathcal{L} : \mathbb{R}^n \to \mathbb{R}$ is a smooth function. A point w^* is a *global* minimizer if $\mathcal{L}(w^*) \leq \mathcal{L}(w)$ for all $w \in \mathbb{R}^n$. Usually $\mathcal{L}$ is a non-convex function and most algorithms are only able to find the *local* minimizer. A point w^* is a local minimizer if there is a neighborhood $\mathcal{N}$ of w^* such that $\mathcal{L}(w^*) \leq \mathcal{L}(w)$ for all $w \in \mathcal{N}$. For convex functions, every local minimizer is also a global minimizer, but this statement is not valid for non-convex functions. If $\mathcal{L}$ is twice continuously differentiable, we may be able to tell that w^* is a local minimizer by examining the gradient $\nabla\mathcal{L}(w^*)$ and the Hessian $\nabla^2\mathcal{L}(w^*)$. Let us assume that the objective function, $\mathcal{L}$, is smooth: the first derivative (gradient) is differentiable and the second derivative (Hessian) is continuous. To study the minimizers of a smooth function, Taylor's theorem is essential.

Theorem 1 (Taylor's Theorem) *Suppose that $\mathcal{L} : \mathbb{R}^n \to \mathbb{R}$ is continuously differentiable. Consider $p \in \mathbb{R}^n$ such that $\mathcal{L}(w + p)$ is well-defined, then we have*

$$\mathcal{L}(w + p) = \mathcal{L}(w) + \nabla\mathcal{L}(w + tp)^T p, \quad \textit{for some } t \in (0, 1). \tag{3}$$

Also if $\mathcal{L}$ is twice continuously differentiable,

$$\mathcal{L}(w + p) = \mathcal{L}(w) + \nabla\mathcal{L}(w + tp)^T p + \frac{1}{2} p^T \nabla^2\mathcal{L}(w + tp) p, \; \textit{for some } t \in (0, 1). \tag{4}$$

A point w^* is a local minimizer of $\mathcal{L}$ only if $\nabla\mathcal{L}(w^*) = 0$. This is known as the *first-order optimality condition.* In addition, if $\nabla^2\mathcal{L}(w^*)$ is positive definite, then

w^* is guaranteed to be a local minimizer. This is known as *second-order sufficient condition* [18].

3 Optimization Strategies

In this section, we briefly introduce two optimization strategies that are commonly used, namely *line search* and *trust-region* methods [18]. Both methods seek to minimize the objective function $\mathcal{L}(w)$ in (1) by defining a sequence of iterates $\{w_k\}$ which are governed by the search direction p_k. Each respective method is defined by its approach to computing the search direction p_k so as to minimize the quadratic model of the objective function defined by

$$\mathcal{Q}_k(p) \triangleq g_k^T p + \frac{1}{2} p^T B_k p, \tag{5}$$

where $g_k \triangleq \nabla \mathcal{L}(w_k)$ and B_k is an approximation to the Hessian matrix $\nabla^2 \mathcal{L}(w_k)$. Note that $\mathcal{Q}_k(p)$ is a quadratic approximation of $\mathcal{L}(w_k + p) - \mathcal{L}(w_k)$ based on the Taylor's expansion in (4).

3.1 Line Search Methods

Each iteration of a line search method computes a search direction p_k by minimizing a quadratic model of the objective function,

$$p_k = \arg\min_{p \in \mathbb{R}^n} \mathcal{Q}_k(p) \triangleq \frac{1}{2} p^T B_k p + g_k^T p, \tag{6}$$

and then decides how far to move along that direction. The iteration is given by $w_{k+1} = w_k + \alpha_k p_k$, where α_k is called the step size. If B_k is a positive definite matrix, the minimizer of the quadratic function can be found as $p_k = -B_k^{-1} g_k$. The ideal choice for step size $\alpha_k > 0$ is the global minimizer of the univariate function $\phi(\alpha) = \mathcal{L}(w_k + \alpha p_k)$, but in practice α_k is chosen to satisfy sufficient decrease and curvature conditions, e.g., the Wolfe conditions [18, 39] given by

$$\mathcal{L}(w_k + \alpha_k p_k) \le \mathcal{L}(w_k) + c_1 \alpha_k \nabla \mathcal{L}(w_k)^T p_k, \tag{7a}$$

$$\nabla \mathcal{L}(w_k + \alpha_k p_k)^T p_k \ge c_2 \nabla \mathcal{L}(w_k)^T p_k, \tag{7b}$$

with $0 < c_1 < c_2 < 1$.

The general pseudocode for the line search method is given in Algorithm 1 (see [18] for details).

Algorithm 1 Line-Search Method.

Input: w_0, tolerance $\epsilon > 0$
$k \leftarrow 0$
repeat
  Compute $g_k = \nabla \mathcal{L}(w_k)$
  Calculate B_k
  Compute search direction p_k by solving (6)
  Find α_k that satisfies Wolfe Conditions in (7b)
  $k \leftarrow k + 1$
until $\|g_k\| < \epsilon$ or k reached to max number of iterations

3.2 *Trust-Region Methods*

Trust-region methods generate a sequence of iterates $w_{k+1} = w_k + p_k$, where each search step, p_k, is obtained by solving the following trust-region subproblem:

$$p_k = \underset{p \in \mathbb{R}^n}{\operatorname{argmin}} \ \mathcal{Q}_k(p) \triangleq \frac{1}{2} p^T B_k p + g_k^T p, \ \text{ such that } \|p\|_2 \leq \delta_k, \tag{8}$$

where $\delta_k > 0$ is the trust-region radius. The global solution to the trust-region subproblem (8) can be characterized by the optimality conditions given in the following theorem due to [40, 41].

Theorem 2 *Let δ_k be a positive constant. A vector p^* is a global solution of the trust-region subproblem* (8) *if and only if $\|p^*\|_2 \leq \delta_k$ and there exists a unique $\sigma^* \geq 0$ such that $B + \sigma^* I$ is positive semi-definite and*

$$(B + \sigma^* I)p^* = -g \quad and \quad \sigma^*(\delta - \|p^*\|_2) = 0. \tag{9}$$

Moreover, if $B + \sigma^* I$ is positive definite, then the global minimizer is unique.

The general pseudocode for the trust-region method is given in Algorithm 2. See Algorithm 6.2 of [18] for details. For further details on trust-region methods, see [42]. See Fig. 1 for an illustration of trust-region methods.

4 Quasi-Newton Optimization Methods

Methods that use $B_k = \nabla^2 \mathcal{L}(w_k)$ for the Hessian in the quadratic model in (5) typically exhibit quadratic rates of convergence. However, in large-scale problems (where n and N are both large), computing the true Hessian explicitly is not practical. In this case, quasi-Newton methods are viable alternatives because they exhibit super-linear convergence rates while maintaining memory and computational efficiency. Instead of the true Hessian, quasi-Newton methods use an approximation, B_k, which

Fig. 1 An illustration of trust-region methods. For indefinite matrices, the Newton step (in red) leads to a saddle-point. The global minimizer (in blue) is characterized by the conditions in Eq. (9) with $B + \sigma^* I$ positive semi-definite. In contrast, local minimizers (in green) satisfy Eq. (9) with $B + \sigma^* I$ not positive semi-definite

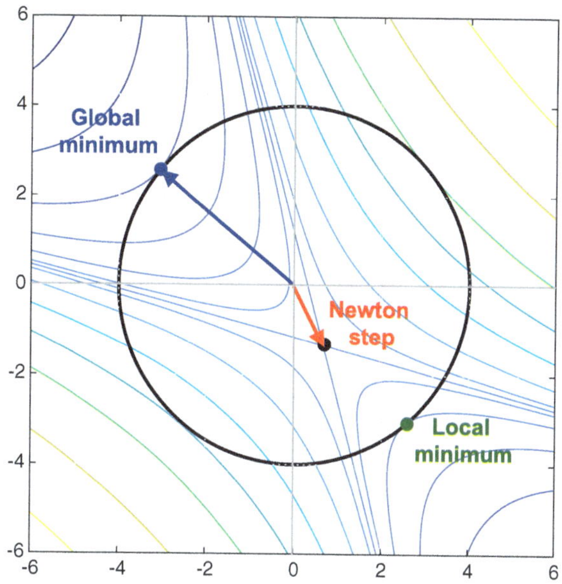

Algorithm 2 Trust-Region Method.

Input: w_0, $\epsilon > 0$, $\hat{\delta} > 0$, $\delta_0 \in (0, \hat{\delta})$, $\eta \in [0, 1/4)$
$k \leftarrow 0$
repeat
  Compute $g_k = \nabla \mathcal{L}(w_k)$
  Construct quasi-Newton matrix B_k
  Compute search direction p_k by solving (8)
  ared $\leftarrow \mathcal{L}(w_k) - \mathcal{L}(w_k + p_k)$
  pred $\leftarrow -\mathcal{Q}_k(p_k)$
  $\rho_k \leftarrow$ ared/pred
  Update trust-region radius δ_k
  if $\rho_k > \eta$ **then**
    $w_{k+1} = w_k + p_k$
  else
    $w_{k+1} = w_k$
  end if
  $k \leftarrow k + 1$
until $\|g_k\| < \epsilon$

is updated after each step to take into account the additional knowledge gained during the step.

Quasi-Newton methods, like gradient descent methods, require only the computation of first-derivative information. They can construct a model of the objective function by measuring the changes in the consecutive gradients for estimating the Hessian. Most methods store the displacement, $s_k \triangleq w_{k+1} - w_k$, and the change of gradients, $y_k \triangleq \nabla \mathcal{L}(w_{k+1}) - \nabla \mathcal{L}(w_k)$, to construct the Hessian approximations, $\{B_k\}$.

The quasi-Newton matrices are required to satisfy the secant equation, $B_{k+1}s_k = y_k$. Typically, there are additional conditions imposed on B_{k+1}, such as symmetry (since the exact Hessian is symmetric), and a requirement that the update to obtain B_{k+1} from B_k is low rank, meaning that the Hessian approximations cannot change too much from one iteration to the next. Quasi-Newton methods vary in how this update is defined. The matrices are defined recursively with the initial matrix, B_0, taken to be $B_0 = \lambda_{k+1} I$, where the scalar $\lambda_{k+1} > 0$.

4.1 The BFGS Update

Perhaps the most well-known among all of the quasi-Newton methods is the Broyden–Fletcher–Goldfarb–Shanno (BFGS) update [18, 43], given by

$$B_{k+1} = B_k - \frac{1}{s_k^T B_k s_k} B_k s_k s_k^T B_k + \frac{1}{y_k^T s_k} y_k y_k^T. \tag{10}$$

The BFGS method generates positive definite approximations whenever the initial approximation $B_0 = \gamma_{k+1} I$ is positive definite and $s_k^T y_k > 0$. A common value for γ_{k+1} is $y_k^T y_k / y_k^T s_k$ [18] (see [44] for alternative methods for choosing γ_{k+1}). Letting

$$S_k \triangleq [s_0 \ \ldots \ s_{k-1}] \text{ and } Y_k \triangleq [y_0 \ \ldots \ y_{k-1}], \tag{11}$$

the BFGS formula can be written in the following compact representation:

$$B_k = B_0 + \Psi_k M_k \Psi_k^T, \tag{12}$$

where Ψ_k and M_k are defined as

$$\Psi_k = \begin{bmatrix} B_0 S_k & Y_k \end{bmatrix}, \quad M_k = \begin{bmatrix} -S_k^T B_0 S_k & -L_k \\ -L_k^T & D_k \end{bmatrix}^{-1}, \tag{13}$$

and L_k is the strictly lower triangular part and D_k is the diagonal part of the matrix $S_k^T Y_k$, i.e., $S_k^T Y_k = L_k + D_k + U_k$, where U_k is a strictly upper triangular matrix. (See [45] for further details.)

It is common in large-scale problems to store only the m most-recently computed pairs $\{(s_k, y_k)\}$, where typically $m \leq 100$. This approach is often referred to as *limited-memory* BFGS (L-BFGS).

4.2 *Line Search L-BFGS Optimization*

In each iteration of the line search (Algorithm 1), we have to compute $p_k = -B_k^{-1} g_k$ at each iteration. We can make use of the following recursive formula for $H_k = B_k^{-1}$:

$$H_{k+1} = \left(I - \frac{y_k s_k^T}{y_k^T s_k} \right) H_k \left(I - \frac{s_k y_k^T}{y_k s_k^T} \right) + \frac{y_k y_k^T}{y_k s_k^T}, \tag{14}$$

where $H_0 = \gamma_{k+1}^{-1} I = {}^{y_k^T y_k}/{}_{y_k^T s_k} I$. The *L-BFGS two-loop recursion algorithm*, given in Algorithm 3, can compute $p_k = -H_k g_k$ in $4mn$ operations [18].

Algorithm 3 L-BFGS Two-Loop Recursion.

$\mathbf{q} \leftarrow g_k = \nabla \mathcal{L}(w_k)$
for $i = k-1, \ldots, k-m$ **do**
  $\alpha_i = \frac{s_i^T q}{y_i^T s_i}$
  $\mathbf{q} \leftarrow \mathbf{q} - \alpha_i y_i$
end for
$\mathbf{r} \leftarrow H_0 q$
for $i = k-1, \ldots, k-m$ **do**
  $\beta = \frac{y_i^T r}{y_i^T s_i}$
  $\mathbf{r} \leftarrow \mathbf{r} + s_i(\alpha_i - \beta)$
end for
return $-\mathbf{r} = -H_k g_k$

4.3 *Trust-Region Subproblem Solution*

To efficiently solve the trust-region subproblem (8), we exploit the compact representation of the BFGS matrix to obtain a global solution based on optimality conditions (9). In particular, we compute the spectral decomposition of B_k using the compact representation of B_k. First, we obtain the QR factorization of $\Psi_k = Q_k R_k$, where Q_k has orthonormal columns and R_k is strictly upper triangular. Then we compute the eigendecomposition of $R_k M_k R_k^T = V_k \hat{\Lambda}_k V_k^T$ so that

$$B_k = B_0 + \Psi_k M_k \Psi_k^T = \gamma_k I + Q_k V_k \hat{\Lambda}_k V_k^T Q_k^T. \tag{15}$$

Note that since V_k is an orthogonal matrix, the matrix $Q_k V_k$ has orthonormal columns. Let $P = [\, Q_k V_k \; (Q_k V_k)^\perp \,] \in \Re^{n \times n}$, where $(Q_k V_k)^\perp$ is a matrix whose columns form an orthonormal basis for the orthogonal complement of the range space of $Q_k V_k$, thereby making P an orthonormal matrix. Then

$$B_k = P \begin{bmatrix} \hat{\Lambda} + \gamma_k I & 0 \\ 0 & \gamma_k I \end{bmatrix} P^T. \tag{16}$$

Using this eigendecomposition to change variables and diagonalize the first optimality condition in (9), a closed form expression for the solution p_k^* can be derived.

The general solution for the trust-region subproblem using the Sherman–Morrison –Woodbury formula is given by

$$p_k^* = -\frac{1}{\tau^*} \left[I - \Psi_k (\tau^* M_k^{-1} + \Psi_k^T \Psi_k)^{-1} \Psi_k^T \right] g_k, \tag{17}$$

where $\tau^* = \gamma_k + \sigma^*$, and σ^* is the optimal Lagrange multiplier in (9) (see [20] for details).

5 Application to Image Recognition

In this section, we compare the line search L-BFGS optimization method with our proposed Trust-Region Minimization Algorithm for Training Responses (TRMinATR). The goal of the experiment is to perform the optimization necessary for neural network training. Both methods are implemented to train the LeNet-5 architecture with the purpose of image classification of the MNIST dataset. All simulations were performed on an AWS EC2 p2.xlarge instance with 1 Tesla K80 GPU, 64 GiB memory, and 4 Intel 2.7 GHz Broadwell processors. For the scalars c_1 and c_2 in the Wolfe line search condition, we used the typical values of $c_1 = 10^{-4}$ and $c_2 = 0.9$ [18]. All code is implemented in the Python language using TensorFlow and it is available at https://rafati.net/lbfgs-tr.

5.1 LeNet-5 Convolutional Neural Network Architecture

We use the convolutional neural network architecture, LeNet-5 (Fig. 2) for computing the likelihood $p_i(y_i|x_i; w_i)$. The LeNet-5 CNN is mainly used in the literature for character and digit recognition tasks [46]. The details of layers' connectivity in LeNet-5 CNN architecture is given in Table 1. The input to the network is 28×28 image and the output is 10 neurons followed by a softmax function, attempting to approximate the posterior probability distribution $p(y_i|x_i; w)$. There are a total of $n = 431{,}080$ trainable parameters (weights) in LeNet-5 CCN.

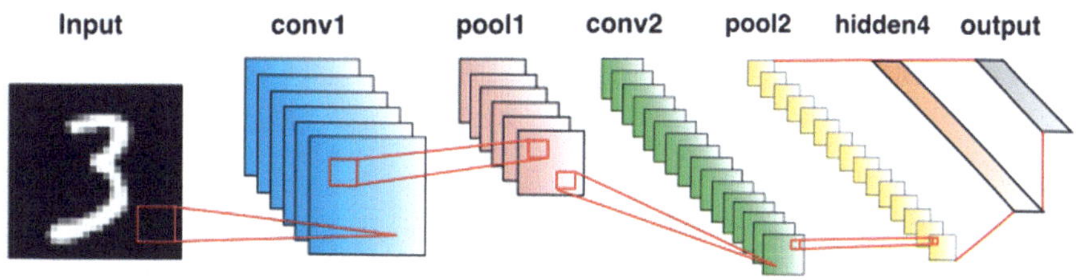

Fig. 2 A LeNet deep learning network inspired by the architecture found in [47]. The neural network is used in the classification of the MNIST dataset of handwritten digits. The convolutional neural network (CNN) uses convolutions followed by pooling layers for feature extraction. The final layer transforms the information into the required probability distribution

Table 1 LeNet-5 CNN architecture [46]

Layer	Connections
0: input	28×28 image
1	convolutional, 20 5×5 filters (stride = 1), followed by ReLU
2	max pooling, 2×2 window (stride = 2)
3	convolutional, 50 5×5 filters (stride = 1), followed by ReLU
4	max pool, 2×2 window (stride = 2)
5	fully connected, 500 neurons (no dropout) followed by ReLU
6: output	fully connected, 10 neurons followed by softmax (no dropout)

5.2 *MNIST Image Classification Task*

The convolutional neural network was trained and tested using the MNIST dataset [48]. The dataset consists of 70,000 examples of handwritten digits with 60,000 examples used as a training set and 10,000 examples used as a test set. The digits range from 0 to 9 and their sizes have been normalized to 28×28 pixel images. The images include labels describing their intended classification. The MNIST dataset consists of 70,000 examples of handwritten image of digits 0–9, with $N = 60{,}000$ image training set $\{(x_i, y_i)\}$, and 10,000 used as the test set. Each image x_i is a 28×28 pixel, and each pixel value is between 0 and 255. Each image x_i in the training set include a label $y_i \in \{0, \ldots, 9\}$ describing its class. The objective function for the classification task in (1) uses the cross entropy between model prediction and true labels given by

$$\ell_i(w) = -\sum_{j=1}^{J} y_{ij} \log(p_i), \tag{18}$$

where the $p_i(x_i; w) = p_i(y = y_i | x_i; w)$ is the probability distribution of the model, i.e., the likelihood that the image is correctly classified, J is the number of classes

($J = 10$ for MNIST digits dataset), and $y_{ij} = 1$ if $j = y_i$ and $y_{ij} = 0$ if $j \neq y_i$ (see [3] for details).

5.3 Results

The line search algorithm and TRMinATR perform comparably in terms of loss and accuracy. This remains consistent with different choices of the memory parameter m (see Fig. 3). The more interesting comparison is that of the training accuracy and the test accuracy. The two metrics follow each other closely. This is unlike the typical results using common gradient descent based optimization. Typically, the test accuracy is delayed in achieving the same results as train accuracy. This would suggest that the model has a better chance of generalizing beyond the training data.

We compare the performance of L-BFGS method with the SGD one. The batch size of 64 was used for training. The loss and accuracy for training batch and the test data has reported for different learning rates, $10^{-6} \leq \alpha \leq 1.0$ (see Fig. 3c–f). Large learning rate (e.g., $\alpha = 1$) led to poor performance (see Fig. 3c, d). The convergence was very slow for the small learning rates, i.e., $\alpha < 10^{-3}$. SGD only succeeds when the proper learning rate was used. This indicates that there is no trivial way for choosing the proper learning rate when using SGD methods and one can run so many experiments for fine-tuning the learning rates. Another interesting observation from Fig. 3 is that L-BFGS method (with either line search or trust-region) has smaller *generalization gap* in comparison to the SGD method. (Generalization gap is the difference between the expected loss and the empirical loss, or roughly the difference between test loss and train loss.)

We also report that the TRMinATR significantly improves the computational efficiency of the line search method when using larger batch sizes. This could be the result of the line search method's need to satisfy certain Wolfe conditions at each iteration. There is also an associated computational cost when verifying that the conditions for sufficient decrease are being met. When the batch size decreases, the trust-region method continues to outperform the line search method. This is especially true when less information is used in the Hessian approximation (see Fig. 4).

6 Application to Deep Reinforcement Learning

6.1 Reinforcement Learning Problem

The reinforcement learning (RL) problem—a class of machine learning—is that of learning through interaction with an *environment*. The learner and decision-maker is called the *agent* and everything outside of the agent is called the *environment*. The agent and environment interact over a sequence of discrete time steps,

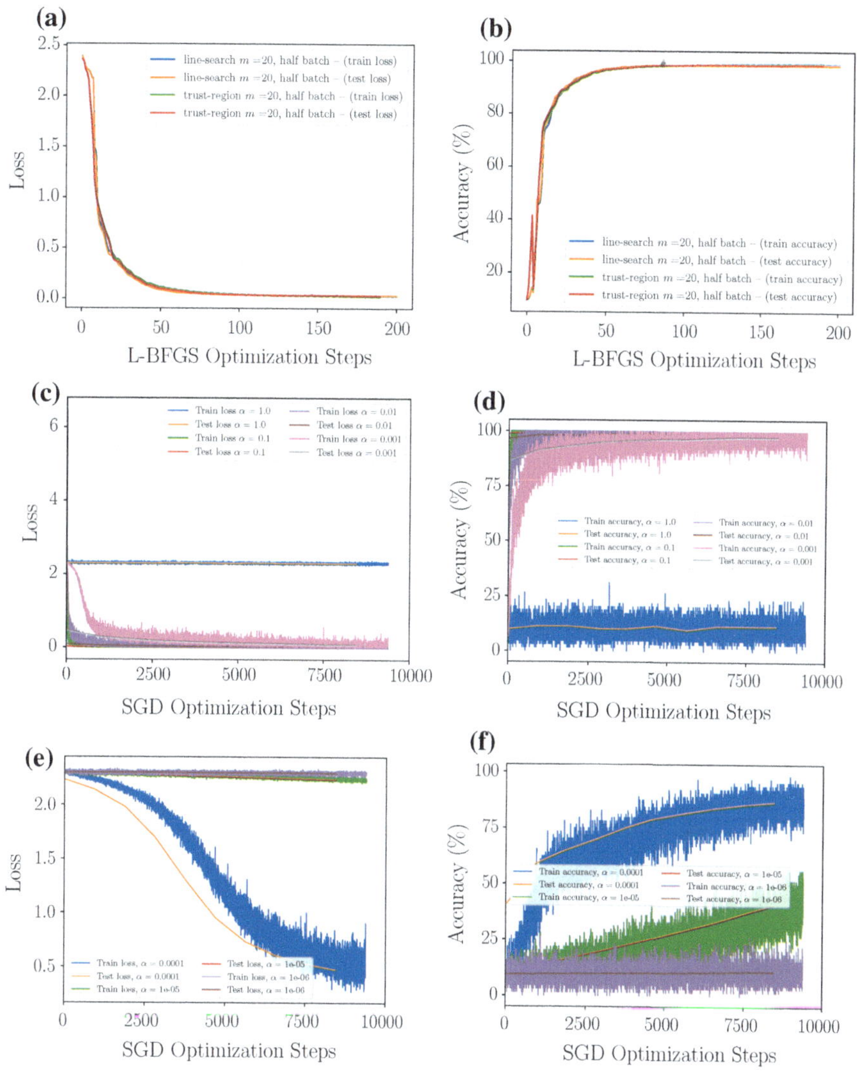

Fig. 3 **a** and **b** Loss and accuracy for the training and test sets, using L-BFGS line search and L-BFGS trust-region methods for $m = 20$ [26]. **c** and **d** Loss and accuracy for the training and test sets using SGD with different learning rates $\alpha \in [1.0, 0.1, 0.01, 0.001]$. **e** and **f** Loss and accuracy for the training and test sets using SGD with small learning rates $\alpha \in [10^{-4}, 10^{-5}, 10^{-6}]$

$t = 0, 1, 2, \ldots, T$. At each time step, t, the agent receives a state, $s_t = s$, from the environment, takes an action, $a_t = a$, and one-time step later, the environment sends a reward, $r_{t+1} = r \in \mathbb{R}$, and an updated state, $s_{t+1} = s'$ (see Fig. 5). Each cycle of interaction, $e = (s, a, r, s')$ is called a transition *experience* (or a trajectory). In an RL problem, the agent should implement a policy, π, from states, $\mathcal{S}$, to possible

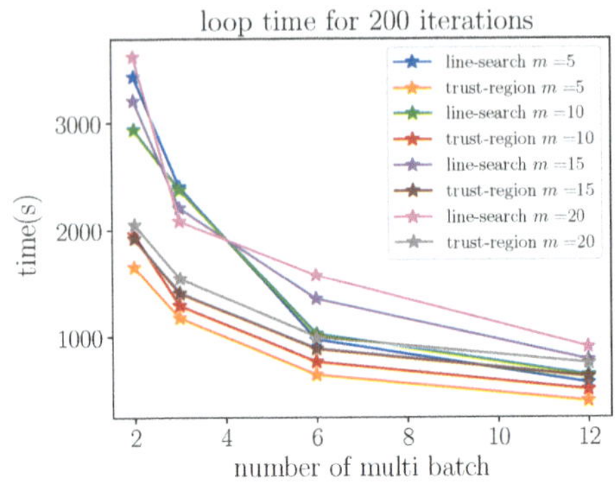

Fig. 4 We compare the loop time for 200 iterations of the line search and trust-region quasi-Newton algorithms for different batch sizes. As the number of multi-batches increases, the size of each batch decreases. Both methods were tested using different values of the memory parameter m [26]

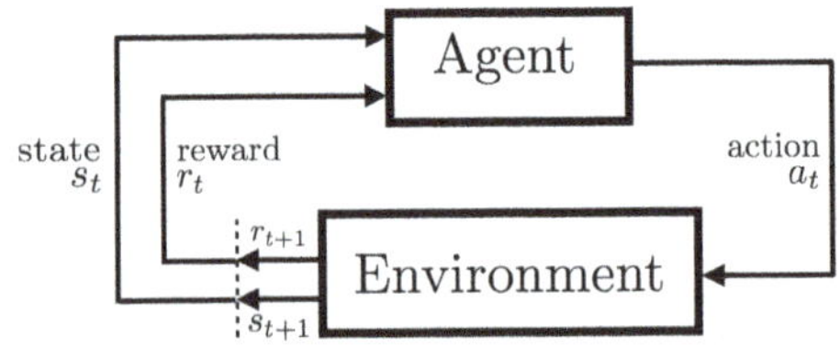

Fig. 5 The agent/environment interaction in reinforcement learning. Adopted from [49]

actions, $\mathcal{A}$. The objective of RL agent is to find an *optimal policy* (best strategy), π^*, that maximizes its expected value of the *return*, G_t, which is the cumulative sum of future rewards from the environment, given by

$$G_t \triangleq r_{t+1} + \gamma r_{t+2} + \gamma^2 r_{t+3} + \cdots = \sum_{t'=t}^{T} \gamma^{t'-t} r_{t'+1}, \tag{19}$$

where $\gamma \in (0, 1]$ is a discount factor and $T \in \mathbb{N}$ is a final step (which can also be infinity) [28]. The optimal policy π^* is defined as

$$\pi^* = \arg\max_{\pi} \mathbb{E}_{\pi}[G_t]. \tag{20}$$

Reinforcement learning is a class of solution methods for solving Markov Decision Processes (MDPs), when the agent does not have prior access to the environment models, i.e., the state transition probabilities, $\mathbb{P}(s'|s, a)$, and the reward function, $\mathcal{R}(s, a)$. Instead, the agent only perceives experiences (or trajectories) from interaction with the environment. The agent can save a limited memory of past experiences (or history) in the set $\mathcal{D}$. It is important to note that each experience, (s, a, s', r), is an example of the joint conditional probability distribution, $p(s', r|s, a)$. Thus the experience memory plays the role of the training data in RL.

It is often useful to define a parametrized value function $Q(s, a; w)$ to estimate the expected value of the return. Q-learning is a model-free RL algorithm that learns a policy without learning a model of the environment. Q-learning algorithm attempts to find the optimal value function by minimizing the expected risk, $L(w)$

$$\min_{w \in \mathbb{R}^n} L(w) \triangleq \frac{1}{2}\mathbb{E}_{(s,a,r,s')\sim p}\Big[\big(\mathcal{Y} - Q(s,a;w)\big)^2\Big], \tag{21}$$

where $\mathcal{Y} = r + \max_{a'} Q(s',a';w)$ is the target value for the expected return based on the *Bellman's optimality* equations [49]. Once the value function is found, the policy function, π, can be found as

$$\pi(s;w) = \arg\max_{a \in \mathcal{A}} Q(s,a;w). \tag{22}$$

6.2 Empirical Risk Minimization in Deep Reinforcement Learning

In practice, the probability distribution over the trajectories, p, is unknown. Therefore, instead of minimizing the expected risk, $L(w)$ in (21), we can define an empirical risk minimization problem over a memory of agent's observed experiences, $\mathcal{D}$, as follows

$$\min_{w \in \mathbb{R}^n} \mathcal{L}(w) \triangleq \frac{1}{2|\mathcal{D}|} \sum_{(s,a,r,s') \in \mathcal{D}} \Big[\big(\mathcal{Y} - Q(s,a;w)\big)^2\Big]. \tag{23}$$

At each optimization step, k, a small set of experiences, J_k, are randomly sampled from the *experience replay memory*, $\mathcal{D}$. This sample is used to compute an stochastic gradient of the objective function, $\nabla\mathcal{L}(w)^{J_k}$, as an approximation for the true gradient, $\nabla\mathcal{L}(w)$,

$$\nabla\mathcal{L}(w)^{(J_k)} \triangleq \frac{-1}{|J_k|} \sum_{e \in J_k} \Big[\big(\mathcal{Y} - Q(s,a;w)\big)\nabla Q\Big]. \tag{24}$$

6.3 L-BFGS Line Search Deep Q-Learning Method

In this section, we propose a novel algorithm for the optimization problem in deep Q-Learning framework, based on the limited-memory BFGS method using a line search strategy. This algorithm is designed to be efficient for parallel computations on GPU. Also the experience memory $\mathcal{D}$ is emptied after each gradient computation, hence the algorithm needs much less RAM memory.

Inspired by [50], we use the overlap between the consecutive multi-batch samples $O_k = J_k \cap J_{k+1}$ to compute y_k as

$$y_k = \nabla \mathcal{L}(w_{k+1})^{(O_k)} - \nabla \mathcal{L}(w_k)^{(O_k)}. \tag{25}$$

The use of overlap to compute y_k has been shown to result in more robust convergence in L-BFGS since L-BFGS uses gradient differences to update the Hessian approximations (see [10, 50]).

At each iteration of optimization, we collect experiences in $\mathcal{D}$ up to batch size b and use the entire experience memory $\mathcal{D}$ as the overlap of consecutive samples O_k. For computing the gradient $g_k = \nabla \mathcal{L}(w_k)$, we use the kth sample, $J_k = O_{k-1} \cup O_k$

$$\nabla \mathcal{L}(w_k)^{(J_k)} = \frac{1}{2}(\nabla \mathcal{L}(w_k)^{(O_{k-1})} + \nabla \mathcal{L}(w_k)^{(O_k)}). \tag{26}$$

Since $\nabla \mathcal{L}(w_k)^{(O_{k-1})}$ is already computed to obtain y_{k-1} in the previous iteration, we only need to compute $\nabla \mathcal{L}^{(O_k)}(w_k)$, given by

$$\nabla \mathcal{L}(w_k)^{(O_k)} = \frac{-1}{|\mathcal{D}|} \sum_{e \in D} \Big[\big(\mathcal{Y} - Q(s, a; w_k) \big) \nabla Q \Big]. \tag{27}$$

Note that in order to obtain y_k, we only need to compute $\nabla \mathcal{L}(w_{k+1})^{(O_k)}$ since $\nabla \mathcal{L}(w_k)^{(O_k)}$ is already computed when we computed the gradient in (26).

The line search multi-batch L-BFGS optimization algorithm for deep Q-Leaning is provided in Algorithm 4.

6.4 Convergence Analysis

In this section, we present a convergence analysis for our deep Q-learning with multi-batch line search L-BFGS optimization method (Algorithm 4). We also provide an analysis of optimality of the state-action value function. We then provide a comparison between the computation time of our deep L-BFGS Q-learning method (Algorithm 4) and that of DeepMind's deep Q-learning algorithm [37], which uses a variant of the SGD method.

6.5 Convergence for the Empirical Risk

To analyze the convergence properties of empirical risk function $\mathcal{L}(w)$ in (23), we assume that

Algorithm 4 Line-Search Multi-Batch L-BFGS Optimization for Deep Q Learning.

Inputs: batch size b, L-BFGS memory m, exploration rate ϵ
Initialize experience memory $\mathcal{D} \leftarrow \emptyset$ with capacity b
Initialize w_0, i.e. parameters of $Q(.,.;w)$ randomly
Initialize optimization iteration $k \leftarrow 0$
for episode $= 1, \ldots, M$ **do**
  Initialize state $s \in \mathcal{S}$
  repeat for each step $t = 1, \ldots, T$
    compute $Q(s, a; w_k)$
    $a \leftarrow$EPS-GREEDY$(Q(s, a; w_k), \epsilon)$
    Take action a
    Observe next state s' and external reward r
    Store transition experience $e = \{s, a, r, s'\}$ to $\mathcal{D}$
    $s \leftarrow s'$
  until s is terminal or intrinsic task is done
  if $|\mathcal{D}| == b$ **then**
    $O_k \leftarrow \mathcal{D}$
    Update w_k by performing **optimization step**
    $\mathcal{D} \leftarrow \emptyset$
  end if
end for

==

Multi-batch line-search L-BFGS Optimization step:
Compute gradient $g_k^{(O_k)}$
Compute gradient $g_k^{(J_k)} \leftarrow \frac{1}{2} g_k^{(O_k)} + \frac{1}{2} g_k^{(O_{k-1})}$
Compute $p_k = -B_k^{-1} g_k^{(J_k)}$ using Algorithm 3
Compute α_k by satisfying the Wolfe Conditions (7b)
Update iterate $w_{k+1} = w_k + \alpha_k p_k$
$s_k \leftarrow w_{k+1} - w_k$
Compute $g_{k+1}^{(O_k)} = \nabla \mathcal{L}(w_{k+1})^{(O_k)}$
$y_k \leftarrow g_{k+1}^{(O_k)} - g_k^{(O_k)}$
Store s_k to S_k and y_k to Y_k and remove oldest pairs
$k \leftarrow k + 1$

$$\mathcal{L}(w) \text{ is strongly convex and twice differentiable.} \tag{28a}$$

$$\text{Given } w, \text{ there are } \lambda, \Lambda > 0 \text{ s.t. } \lambda I \preceq \nabla^2 \mathcal{L}(w) \preceq \Lambda I, \tag{28b}$$

$$\text{Given } w, \text{ there is } \eta > 0 \text{ such that } \|\nabla \mathcal{L}(w)\|^2 \leq \eta^2. \tag{28c}$$

In (28b) we assume that the eigenvalues of the Hessian matrix are bounded, and in (28c) we assume that gradients are not unbounded.

Lemma 1 *Given w, there exist* $\lambda', \Lambda' > 0$ *such that* $\lambda' I \preceq H_k \preceq \Lambda' I$.

Proof Due to the assumptions (28a) and (28b), the eigenvalues of the positive definite matrix H_k are also bounded [50, 51]. □

Lemma 2 *Let* w^* *be a minimizer of* $\mathcal{L}$. *Then, for all w, we have* $2\lambda(\mathcal{L}(w) - \mathcal{L}(w^*) \leq \|\nabla \mathcal{L}(w)\|^2$.

Proof For any convex function, $\mathcal{L}$ and for any two points, w and w^*, one can show that

$$\mathcal{L}(w) \leq \mathcal{L}(w^*) + \nabla\mathcal{L}(w^*)^T(w - w^*) + \frac{1}{2\lambda}\|\nabla\mathcal{L}(w) - \nabla\mathcal{L}(w^*)\|^2. \tag{29}$$

(see [52]). Since w^* is a minimizer of $\mathcal{L}$, $\nabla\mathcal{L}(w^*) = 0$ in (29), which completes the proof. □

Theorem 3 *Let w_k be iterates generated by Algorithm 4, and assume that the step length, α_k, is fixed. The upper bound for the empirical risk offset from the true minimum value is*

$$\begin{aligned}\mathcal{L}(w_k) - \mathcal{L}(w^*) &\leq (1 - 2\alpha\lambda\lambda')^k\big[\mathcal{L}(w_0) - \mathcal{L}(w^*)\big] \\ &\quad + \big[1 - (1 - 2\alpha\lambda\lambda')^k\big]\frac{\alpha^2\Lambda'^2\Lambda\eta^2}{4\lambda'\lambda}.\end{aligned} \tag{30}$$

Proof Using the Taylor expansion of

$$\mathcal{L}(w_{k+1}) = \mathcal{L}(w_k - \alpha_k H\nabla\mathcal{L}(w_k))$$

around w_k, we have

$$\mathcal{L}(w_{k+1}) \leq \mathcal{L}(w_k) - \alpha_k\nabla\mathcal{L}(w_k)^T H_k\nabla\mathcal{L}(w_k) + \frac{\Lambda}{2}\|\alpha_k\nabla\mathcal{L}(w_k)^T H_k\nabla\mathcal{L}(w_k)\|^2. \tag{31}$$

By applying assumptions (28) and Lemmas 1 and 2 to the above inequality, we have

$$\mathcal{L}(w_{k+1}) \leq \mathcal{L}(w_k) - 2\alpha_k\lambda'\lambda[\mathcal{L}(w_k) - \mathcal{L}(w^*)] + \frac{\alpha_k^2\Lambda'^2\Lambda\eta^2}{4\lambda'\lambda}. \tag{32}$$

By rearranging terms and using recursion expression and recursion over k, we have the proof. For a more detailed proof, see [50, 51]. □

If the step size is bounded, $\alpha \in (0, 1/2\lambda\lambda')$, we can conclude that the first term of the bound given in (30) is decaying linearly to zero when $k \to \infty$ and the constant residual term, $\frac{\alpha^2\Lambda'^2\Lambda\eta^2}{4\lambda'\lambda}$, is the neighborhood of convergence.

6.6 Value Optimality

The Q-learning method has been proved to converge to the optimal value function if the step sizes satisfies $\sum_k \alpha_k = \infty$ and $\sum_k \alpha_k^2 < \infty$ [53]. Now, we want to prove that Q-learning using the L-BFGS update also theoretically converges to the optimal value function under one additional condition on the step length, α_k.

Theorem 4 *Let Q^* be the optimal state-action value function and Q_k be the Q-function with parameters w_k. Furthermore, assume that the gradient of Q is bounded, $\|\nabla Q\|^2 \leq \eta''^2$, and the Hessian of Q functions satisfy $\lambda'' \preceq \nabla^2 Q \preceq \Lambda''$. We have*

$$\|Q_{k+1} - Q^*\|_\infty < \prod_{j=0}^{k} \Big[1 - \alpha_j \eta''^2 \lambda + \frac{\alpha_j \eta''^2 \Lambda'^2 \Lambda''}{2}\Big]^k \|Q_0 - Q^*\|_\infty. \tag{33}$$

If step size α_k satisfies

$$\Big|1 - \alpha_k \eta''^2 \lambda + \frac{\alpha_k \eta \eta' \Lambda'^2 \Lambda''}{2}\Big| \leq \mu < 1, \quad \textit{for all } k, \tag{34}$$

$Q(.,.; w_k)$ ultimately will converge to Q^*, as $k \to \infty$.

Proof First we derive the effect of the parameter update from w_k to

$$w_{k+1} = w_k - \alpha_k H_k \nabla \mathcal{L}(w_k)$$

on the optimality neighbor.

$$\|Q_{k+1} - Q^*\|_\infty \triangleq \max_{s,a} \big|Q(s, a, w_{k+1}) - Q^*(s, a)\big| \tag{35}$$

We approximate the gradient using only one experience (s, a, r, s'),

$$\nabla \mathcal{L}(w_k) \approx \big(Q(s, a; w_k) - Q^*(s, a; w_k)\big)\nabla Q_k(s, a; w_k). \tag{36}$$

Using Taylor's expansion to approximate $Q(s, a, w_{k+1})$ results in

$$\begin{aligned} &Q(s, a; w_{k+1}) = Q(s, a; w_k - \alpha_k H_k \nabla \mathcal{L}(w_k)) \\ &= Q(s, a; w_k) - \alpha_k \nabla \mathcal{L}_k^T H_k \nabla Q_k + \frac{\alpha_k^2}{2} \nabla \mathcal{L}_k^T H_k \nabla^2 Q(\xi_k) H_k \nabla \mathcal{L}_k^T \\ &= Q_k - \alpha_k (Q_k - Q^*) \nabla Q_k^T H_k \nabla Q_k + \frac{\alpha_k^2}{2} (Q_k - Q^*) \nabla Q_k^T H_k \nabla^2 Q(\xi_k) H_k \nabla \mathcal{L}_k^T, \end{aligned} \tag{37}$$

where ξ is between w_k and w_{k+1}, $Q_k := Q(s, a; w_k)$, $\nabla Q_k := \nabla Q(s, a; w_k)$, and $\nabla \mathcal{L}_k := \nabla \mathcal{L}(w_k)$. We can use the above expression to compute $\|Q_{k+1} - Q^*\|_\infty$:

$$\begin{aligned} &\|Q_{k+1} - Q^*\|_\infty = \\ &\max_{s,a} \Big|(Q_k - Q^*)\Big[1 - \alpha_k \nabla Q_k^T H_k \nabla Q_k + \frac{\alpha_k^2}{2} \nabla Q_k^T H_k \nabla^2 Q(\xi_k) H_k \nabla \mathcal{L}_k\Big]\Big|_{(s,a)}. \end{aligned} \tag{38}$$

If α_k satisfies

$$\left|1 - \alpha_k \nabla Q_k^T H_k \nabla Q_k + \frac{\alpha_k^2}{2} \nabla Q_k^T H_k \nabla^2 Q_k H_k \nabla \mathcal{L}_k\right| \leq \mu < 1, \tag{39}$$

then

$$\|Q_{k+1} - Q^*\|_\infty \leq \mu \|Q_k - Q^*\|_\infty \leq \mu^{k+1} \|Q_0 - Q^*\|_\infty. \tag{40}$$

Therefore, Q_k converges to Q^* when $k \to \infty$. Considering our assumptions on the bounds of the eigenvalues of $\nabla^2 Q_k$ and H_k, we can derive (34) from (39). Recursion on (38) from $k = 0$ to $k + 1$ results in (33). □

6.7 *Computation Time*

Let us compare the cost of deep L-BFGS Q-learning in Algorithm 4 with DQN algorithm in [37] that uses a variant of SGD. Assume that the cost of computing gradient is $O(bn)$ where b is the batch size. The real cost is probably less than this due to the parallel computation on GPUs. Let us assume that we run both algorithm for L steps. We update the weights every b steps. Hence there is L/b maximum updates in our algorithm. The SGD batch size in [37], b_s, is smaller than b, but the frequency of the update is high, $f \ll b$. Each iteration of the L-BFGS algorithm update introduces the cost of computing the gradient, $g_k^{(O_k)}$, which is $O(bn)$, the cost of computing the search step, $p_k = -H_k g_k^{(O_k)}$, using L-BFGS two-loop recursion (Algorithm 3), which is $O(4mn)$, and the cost of satisfying the Wolfe conditions (7b) to find a step size that usually satisfies for $\alpha = 1$ and, in some steps, requires recomputing the gradient z times. Therefore we have

$$\frac{\text{Cost of Algorithm 3}}{\text{Cost of DQN [37]}} = \frac{(L/b)(zbn + 4mn)}{(L/f)(b_s n)} = \frac{fz}{b_s} + \frac{4fm}{bb_s}. \tag{41}$$

In our algorithm, we use a quite large batch size to compute less noisy gradients. With $b = 2048, b_s = 32, f = 4, z = 5, m = 20$, the runtime cost ratio will be around $0.63 < 1$. Although the per-iteration cost of the the SGD algorithm is lower than L-BFGS, the total training time of our algorithm is less than DQN [37] for the same number of RL steps due to the need for less frequent updates in the L-BFGS method.

6.8 *Experiments on Atari 2600 Games*

We performed experiments using our approach (Algorithm 4) on six Atari 2600 games—Beam Rider, Breakout, Enduro, Q*bert, Seaquest, and Space Invaders. We used OpenAI's gym Atari environments [54] which are wrappers on the Arcade Learning Environment emulator [55]. These games have been used by other researchers investigating different learning methods [37, 55–58], and, hence, they serve as benchmark environments for the evaluation of deep reinforcement learning algorithms.

We used DeepMind's deep Q-Network (DQN) architecture, described in [37], as a function approximator for $Q(s, a; w)$. The same architecture was used to train agents to play different Atari games. The raw Atari frames, which are 210×160 pixel images with a 128 color palette, were preprocessed by first converting their RGB representation to grayscale and then downsampling the images to be 110×84 pixels. The final input representation is obtained by cropping an 84×84 region of the image that roughly captures the playing area. The stack of the last four consecutive frames was used to produce the input, of size $(4 \times 84 \times 84)$, to the Q-function. The first hidden layer of the network consisted of 32 convolutional filters of size 8×8 with stride 4, followed by a Rectified Linear Unit (ReLU) for nonlinearity. The second hidden layer consisted of 64 convolutional filters of size 4×4 with stride 2, followed by a ReLU function. The third layer consisted of 512 fully connected linear units, followed by ReLU. The output layer was a fully connected linear layer with an output, $Q(s, a_i, w)$, for each valid joystick action, $a_i \in \mathcal{A}$. The number of valid joysticks actions, i.e., $|\mathcal{A}|$, was 9 for Beam Rider, 4 for Breakout, 9 for Enduro, 6 for Q*Bert, 18 for Seaquest, and 6 for Space Invaders.

We only used 2 million (2000×1024) Q-learning training steps for training the network on each game (instead of 50 million steps that were used originally in [37]). The training was stopped when the norm of the gradient, $\|g_k\|$, was less than a threshold. We used ϵ-greedy for an exploration strategy, and, similar to [37], the exploration rate, ϵ, was annealed linearly from 1 to 0.1.

Every 10,000 steps, the performance of the learning algorithm was tested by freezing the Q-network's parameters. During the test time, we used $\epsilon = 0.05$. The greedy action, $\max_a Q(s, a; w)$, was chosen by the Q-network 95% of the times and there was 5% randomness, similar to the DeepMind implementation in [37].

Inspired by [37], we also used separate networks to compute the target values, $\mathcal{Y} = r + \gamma \max_{a'} Q(s', a', w_{k-1})$, which was essentially the network with parameters in previous iterate. After each iteration of the multi-batch line search L-BFGS, w_k was updated to w_{k+1}, and the target network's parameter w_{k-1} was updated to w_k.

Our optimization method was different than DeepMind's RMSProp method, used in [37] (which is a variant of SGD). We used a stochastic line search L-BFGS method as the optimization method (Algorithm 4). There are a few important differences between our implementation of deep reinforcement learning and DeepMind's DQN algorithm.

We used a quite large batch size, b, in comparison to [37]. We experimented with different batch sizes $b \in \{512, 1024, 2048, 4096, 8192\}$. The experience memory, $\mathcal{D}$, had a capacity of b also. We used one NVIDIA Tesla K40 GPU with 12GB GDDR5 RAM. The entire experience memory, $\mathcal{D}$, could fit in the GPU RAM with a batch size of $b \leq 8192$.

After every b steps of interaction with the environment, the optimization step in Algorithm 4 was run. We used the entire experience memory, $\mathcal{D}$, for the overlap, O_k, between two consecutive samples, J_k and J_{k+1}, to compute the gradient in (27) as well as y_k in (25). Although the DQN algorithm used a smaller batch size of 32, the frequency of optimization steps was high (every 4 steps). We hypothesize that using the smaller batch size made the computation of the gradient too noisy, and, also, this approach does not save significant computational time since the overhead of data transfer between GPU and CPU is more costly than the computation of the gradient over a bigger batch size, due to the power of parallelism in a GPU. Once the overlap gradient, $g_k^{(O_k)}$, was computed, we computed the gradient, $g_k^{(J_k)}$, for the current sample, J_k, in (26) by memorizing and using the gradient information from the previous optimization step. Then, the L-BFGS two-loop recursion in Algorithm 3 was used to compute the search direction $p_k = -H_k g_k^{(J_k)}$.

After finding the quasi-Newton decent direction, p_k, the Wolfe Condition (7b) was applied to compute the step size, $\alpha_k \in [0.1, 1]$, by satisfying the sufficient decrease and the curvature conditions [18, 39]. During the optimization steps, either the step size of $\alpha_k = 1$ satisfied the Wolfe conditions, or the line search algorithm iteratively used smaller α_k until it satisfied the Wolfe conditions or reached a lower bound of 0.1. The original DQN algorithm used a small fixed learning rate of 0.00025 to avoid the execrable drawback of the noisy stochastic gradient descent step, which makes the learning process very slow.

The vectors $s_k = w_{k+1} - w_k$ and $y_k = g_{k+1}^{(O_k)} - g_k^{(O_k)}$ were only added to the recent collections S_k and Y_k if $s_k^T y_k > 0$ and not close to zero. We applied this condition to *cautiously* preserve the positive definiteness of the L-BFGS matrices B_k. Only the m recent $\{(s_i, y_i)\}$ pairs were stored into S_k and Y_k ($|S_k| = m$ and $|Y_k| = m$) and the older pairs were removed from the collections. We experimented with different L-BFGS memory sizes $m \in \{20, 40, 80\}$.

All code is implemented in the Python language using Pytorch, NumPy, and SciPy libraries, and it is available at http://rafati.net/quasi-newton-rl.

6.9 Results and Discussion

The average of the maximum game scores is reported in Fig. 6a. The error bar in Fig. 6a is the standard deviation for the simulations with different batch size, $b \in \{512, 1024, 2048, 4096\}$, and different L-BFGS memory size, $m \in \{20, 40, 80\}$, for each Atari game (total of 12 simulations per each task). All simulations regardless of the batch size, b, and the memory size, m, exhibited robust learning. The average training time for each task, along with the empirical loss values, $\mathcal{L}(w_k)$, is shown in Fig. 6b.

The Coefficient of Variation (CV) for the test scores was about 10% for each Atari task. (The coefficient of variation is defined as the standard deviation divided by the mean). We did not find a correlation between the test scores and the different batch sizes, b, or the different L-BFGS memory sizes, m. The coefficient of variation for the training times was about 50% for each Atari task. Hence, we did not find a strong correlation between the training time and the different batch sizes, b, or the different L-BFGS memory sizes, m. In most of the simulations, the loss for the training time, as shown in Fig. 6b, was very small.

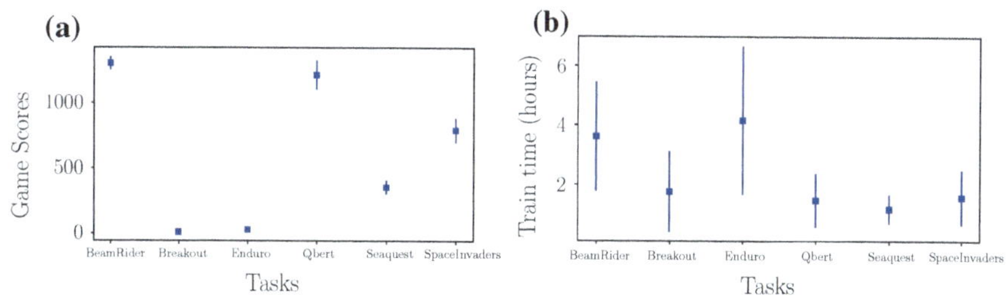

Fig. 6 **a** Test scores **b** Total training time for Atari games

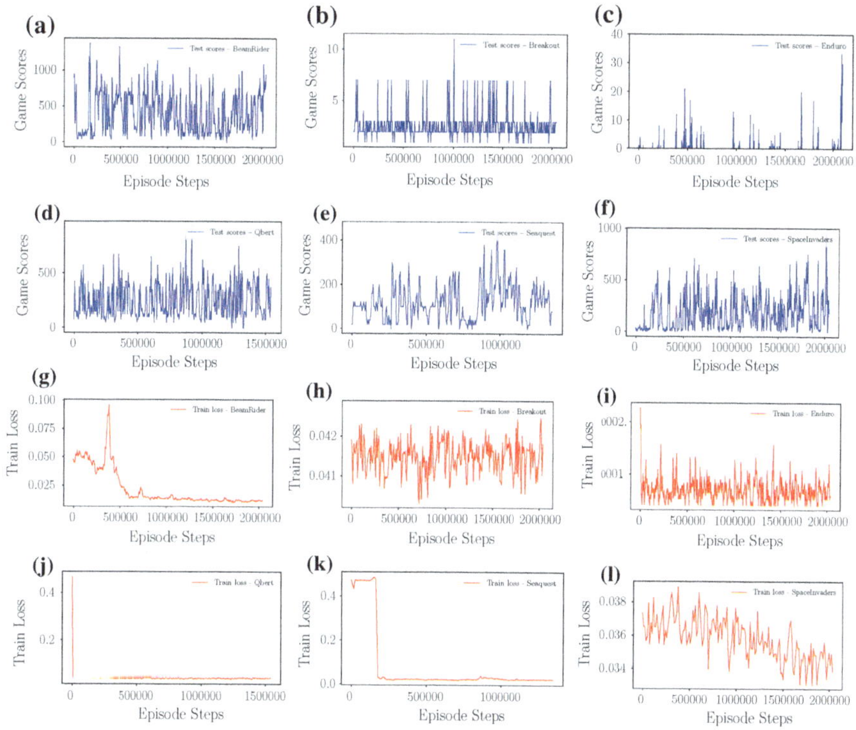

Fig. 7 **a–f** Test scores and **g–l** Training loss for six Atari games—Beam Rider, Breakout, Enduro, Q*bert, Seaquest, and Space Invaders. The results are form simulations with batch size $b = 2048$ and the L-BFGS memory size $m = 40$

The test scores and the training loss, $\mathcal{L}_k$, for the six Atari 2600 environments is shown in Fig. 7 using the batch size of $b = 2048$ and L-BFGS memory size $m = 40$.

The results of the deep L-BFGS Q-Learning algorithm is summarized in Table 2, which also includes an expert human performance and some recent model-free methods: the Sarsa algorithm [55], the *contingency aware* method from [56], deep Q-learning [36], and two methods based on policy optimization called Trust-Region

Table 2 Best game scores for Atari 2600 games with different learning methods. Beam Rider, Breakout, Enduro, Q*bert, Seaquest, and Space Invaders

Method	Beam Rider	Breakout	Enduro	Q*bert	Seaquest	Space Invaders
Random	354	1.2	0	157	110	179
Human	7456	31	368	18900	28010	3690
Sarsa [55]	996	5.2	129	614	665	271
Contingency [56]	1743	6	159	960	723	268
HNeat Pixel [57]	1332	4	91	1325	800	1145
DQN [36]	4092	168	470	1952	1705	581
TRPO, Single path [58]	1425	10	534	1973	1908	568
TRPO, Vine [58]	859	34	431	7732	7788	450
SGD ($\alpha = 0.01$)	804	13	2	1325	420	735
SGD ($\alpha = 0.00001$)	1092	14	1	1300	380	975
Our method	1380	18	49	1525	600	955

Policy Optimization (TRPO vine and TRPO single path) [58] and the Q-learning with the SGD method. Our method outperformed most other methods in the Space Invaders game. Our deep L-BFGS Q-learning method consistently achieved reasonable scores in the other games. Our simulations were only trained on about 2 million Q-learning steps (much less than other methods). DeepMind DQN method outperformed our algorithm on most of the games, except on the Space Invaders game.

The training time for our simulations was on the order of 3 h (4 h for Beam Rider, 2 h for Breakout, 4 h for Enduro, 2 h for Q*bert, 1 h for Seaquest, and 2 h for Space Invaders). Our method outperformed all other methods of computational time. For example, 500 iterations of the TRPO algorithm took about 30 h [58]. We also compared our method with the SGD method. For each task, we trained the Q-learning algorithm using the SGD optimization method for two million Q-learning training steps. We examined two different learning rates: a relatively large learning rate, $\alpha = 0.01$, and a very small learning rate, $\alpha = 0.00001$. The other parameters were adopted from [37]. The game scores with our method outperformed the SGD method in most of the simulations (11 out of 12 times) (see Table 2). Although the computation time per iteration of the SGD update is lower than our method, but the total training time of the SGD method is much slower than our method due to the higher frequency of the parameter updates in the SGD method as opposed to our L-BFGS line search method. See Table 3 for the results of the training time for each task, using different optimization methods, L-BFGS and SGD.

Table 3 Average training time for Atari 2600 games with different learning methods (in hours). Beam Rider, Breakout, Enduro, Q*bert, Seaquest, and Space Invaders

Method	Beam Rider	Breakout	Enduro	Q*bert	Seaquest	Space Invaders
SGD ($\alpha = 0.01$)	4	2	8	8	8	1
SGD ($\alpha = 0.00001$)	7	11	7	6	8	6
Our method	4	2	4	2	1	1

7 Conclusions

In this chapter, we implemented an optimization method based on the limited-memory quasi-Newton method known as L-BFGS as an alternative to the gradient descent methods typically used to train deep neural networks. We considered both line search and trust-region frameworks. The contribution of this research is an algorithm known as TRMinATR which minimizes the cost function of the neural network by efficiently solving a sequence of trust-region subproblems using low-rank updates to Hessian approximations. The benefit of the method is that the algorithm is free from the constraints of data-specific parameters seen in traditionally used methods. TRMinATR also improves the computational efficiency of a similar line search implementation. Furthermore, we proposed and implemented a novel optimization method based on line search limited-memory BFGS for the deep reinforcement learning framework. We tested our method on six classic Atari 2600 games. The L-BFGS method attempts to approximate the Hessian matrix by constructing positive definite matrices with low-rank updates. Due to the nonconvex and nonlinear loss functions arising in deep reinforcement learning, our numerical experiments show that using the curvature information in computing the search direction leads to a more robust convergence when compared to the SGD results. Our proposed deep L-BFGS Q-Learning method is designed to be efficient for parallel computations on GPUs. Our method is much faster than the existing methods in the literature, and it is memory efficient since it does not need to store a large experience replay memory. Since our proposed limited-memory quasi-Newton optimization methods rely only on first-order gradients, they can be efficiently scaled and employed for larger scale supervised learning, unsupervised learning, and reinforcement learning applications. The overall enhanced performance of our proposed optimization methods on deep learning applications can be attributed to the robust convergence properties, fast training time, and better generalization characteristics of these optimization methods.

References

1. I. Goodfellow, Y. Bengio, A. Courville, *Deep Learning* (MIT Press, Cambridge, 2016)
2. M.A. Wani, F.A. Bhat, S. Afzal, A. Khan, *Advances in Deep Learning* (Springer, Berlin, 2020)
3. T. Hastie, R. Tibshirani, J. Friedman, *The Elements of Statistical Learning: Data Mining, Inference and Prediction*, 2nd edn. (Springer, Berlin, 2009)
4. H. Robbins, S. Monro, A stochastic approximation method. Ann. Math. Stat. **22**(3), 400–407 (1951)
5. L. Bottou, Large-scale machine learning with stochastic gradient descent, in *Proceedings of COMPSTAT'2010* (Springer, 2010), pp. 177–186
6. J.C. Duchi, E. Hazan, Y. Singer, Adaptive subgradient methods for online learning and stochastic optimization. J. Mach. Learn. Res. **12**, 2121–2159 (2011)
7. B. Recht, C. Re, S. Wright, F. Niu, Hogwild: a lock-free approach to parallelizing stochastic gradient descent, in *Advances in Neural Information Processing Systems* (2011), pp. 693–701
8. L. Adhikari, O. DeGuchy, J.B. Erway, S. Lockhart, R.F. Marcia, Limited-memory trust-region methods for sparse relaxation, in *Wavelets and Sparsity XVII*, vol. 10394 (International Society for Optics and Photonics, 2017)
9. Q.V. Le, J. Ngiam, A. Coates, A. Lahiri, B. Prochnow, A.Y. Ng, On optimization methods for deep learning, in *Proceedings of the 28th International Conference on International Conference on Machine Learning* (2011), pp. 265–272
10. J.B. Erway, J. Griffin, R.F. Marcia, R. Omheni, Trust-region algorithms for training responses: machine learning methods using indefinite Hessian approximations. Optim. Methods Softw. 1–28 (2019)
11. P. Xu, F. Roosta-Khorasan, M.W. Mahoney, Second-order optimization for non-convex machine learning: an empirical study. ArXiv e-prints (2017)
12. J. Martens, Deep learning via Hessian-free optimization, in *Proceedings of the 27th International Conference on Machine Learning (ICML)* (2010), pp. 735–742
13. J. Martens, I. Sutskever, Learning recurrent neural networks with hessian-free optimization, in *Proceedings of the 28th International Conference of on Machine Learning (ICML)* (2011), pp. 1033–1040
14. J. Martens, I. Sutskever, Training deep and recurrent networks with hessian-free optimization, in *Neural Networks: Tricks of the Trade* (Springer, 2012), pp. 479–535
15. R. Bollapragada, R.H. Byrd, J. Nocedal, Exact and inexact subsampled newton methods for optimization. IMA J. Numer. Anal. **39**(2), 545–578 (2018)
16. M.D. Zeiler, ADADELTA: an adaptive learning rate method (2012). arxiv:1212.5701
17. D.P. Kingma, J. Ba, Adam: a method for stochastic optimization (2014). arXiv:1412.6980
18. J. Nocedal, S.J. Wright, *Numerical Optimization*, 2nd edn. (Springer, New York, 2006)
19. J. Brust, O. Burdakov, J.B. Erway, R.F. Marcia, A dense initialization for limited-memory quasi-newton methods. Comput. Optim. Appl. **74**(1), 121–142 (2019)
20. J. Brust, J.B. Erway, R.F. Marcia, On solving L-SR1 trust-region subproblems. Comput. Optim. Appl. **66**(2), 245–266 (2017)
21. C.G. Broyden, The convergence of a class of double-rank minimization algorithms 1. General considerations. SIAM J. Appl. Math. **6**(1), 76–90 (1970)
22. R. Fletcher, A new approach to variable metric algorithms. Comput. J. **13**(3), 317–322 (1970)
23. D. Goldfarb, A family of variable-metric methods derived by variational means. Math. Comput. **24**(109), 23–26 (1970)
24. D.F. Shanno, Conditioning of quasi-Newton methods for function minimization. Math. Comput. **24**(111), 647–656 (1970)
25. Q.V. Le, J. Ngiam, A. Coates, A. Lahiri, B. Prochnow, A.Y. Ng, On optimization methods for deep learning, in *Proceedings of the 28th International Conference on International Conference on Machine Learning* (Omnipress, 2011), pp. 265–272
26. J. Rafati, O. DeGuchy, R.F. Marcia, *Trust-region minimization algorithm for training responses (TRMinATR): the rise of machine learning techniques, in 26th European Signal Processing Conference (EUSIPCO 2018)* (Italy, Rome, 2018)

27. O. Burdakov, L. Gong, Y.X. Yuan, S. Zikrin, On efficiently combining limited memory and trust-region techniques. Math. Program. Comput. **9**, 101–134 (2016)
28. R.S. Sutton, A.G. Barto, *Reinforcement Learning: An Introduction*, 2nd edn. (MIT Press, Cambridge, 2018)
29. R.S. Sutton, Generalization in reinforcement learning: successful examples using sparse coarse coding. Adv. Neural Inf. Process. Syst. **8**, 1038–1044 (1996)
30. J. Rafati, D.C. Noelle, *Lateral inhibition overcomes limits of temporal difference learning, in Proceedings of the 37th Annual Cognitive Science Society Meeting* (Pasadena, CA, USA, 2015)
31. J. Rafati, D.C. Noelle, *Sparse coding of learned state representations in reinforcement learning, in Conference on Cognitive Computational Neuroscience* (New York City, NY, USA, 2017)
32. J. Rafati Heravi, Learning representations in reinforcement learning. Ph.D. Thesis, University of California, Merced, 2019
33. J. Rafati, D.C. Noelle, Learning representations in model-free hierarchical reinforcement learning (2019). arXiv:1810.10096
34. F.S. Melo, S.P. Meyn, M.I. Ribeiro, An analysis of reinforcement learning with function approximation, in *Proceedings of the 25th International Conference on Machine Learning* (2008)
35. G. Tesauro, Temporal difference learning and TD-Gammon. Commun. ACM **38**(3)(1995)
36. V. Mnih, K. Kavukcuoglu, D. Silver, A. Graves, I. Antonoglou, D. Wierstra, M.A. Riedmiller, Playing Atari with deep reinforcement learning (2013). arxiv:1312.5602
37. V. Mnih, K. Kavukcuoglu, D. Silver, Others, Human-level control through deep reinforcement learning. Nature **518**(7540), 529–533 (2015)
38. D. Silver, A. Huang, C.J. Maddison, D. Hassabis, Others, Mastering the game of go with deep neural networks and tree search. Nature **529**(7587), 484–489 (2016)
39. P. Wolfe, Convergence conditions for ascent methods. SIAM Rev. **11**(2), 226–235 (1969)
40. D.M. Gay, Computing optimal locally constrained steps. SIAM J. Sci. Stat. Comput. **2**(2), 186–197 (1981)
41. J.J. Moré, D.C. Sorensen, Computing a trust region step. SIAM J. Sci. Stat. Comput. **4**(3), 553–572 (1983)
42. A.R. Conn, N.I.M. Gould, P.L. Toint, *Trust-Region Methods* (Society for Industrial and Applied Mathematics, Philadelphia, 2000)
43. D.C. Liu, J. Nocedal, On the limited memory BFGS method for large scale optimization. Math. Program. **45**(1–3), 503–528 (1989)
44. J. Rafati, R.F. Marcia, *Improving L-BFGS initialization for trust-region methods in deep learning, in 17th IEEE International Conference on Machine Learning and Applications* (Orlando, Florida, 2018)
45. R.H. Byrd, J. Nocedal, R.B. Schnabel, Representations of quasi-Newton matrices and their use in limited-memory methods. Math. Program. **63**, 129–156 (1994)
46. Y. LeCun, L. Bottou, Y. Bengio, P. Haffner, Gradient-based learning applied to document recognition. Proc. IEEE **86**(11), 2278–2324 (1998)
47. Y. LeCun, Others, Lenet5, convolutional neural networks (2015), p. 20
48. Y. LeCun, The MNIST database of handwritten digits (1998). http://yann.lecun.com/exdb/mnist/
49. R.S. Sutton, A.G. Barto, *Reinforcement Learning: An Introduction*, 1st edn. (MIT Press, Cambridge, 1998)
50. A.S. Berahas, J. Nocedal, M. Takac, A multi-batch L-BFGS method for machine learning, in *Advances in Neural Information Processing Systems*, vol. 29 (2016), pp. 1055–1063
51. R.H. Byrd, S.L. Hansen, J. Nocedal, Y. Singer, A stochastic quasi-newton method for large-scale optimization. SIAM J. Optim. **26**(2), 1008–1031 (2016)
52. Y. Nesterov, *Introductory Lectures on Convex Optimization: A Basic Course* (Springer Science & Business Media, Berlin, 2013)
53. T. Jaakkola, M.I. Jordan, S.P. Singh, On the convergence of stochastic iterative dynamic programming algorithms. Neural Comput. **6**(6), 1185–1201 (1994)
54. G. Brockman, V. Cheung, L. Pettersson, J. Schneider, J. Schulman, J. Tang, W. Zaremba, OpenAI Gym (2016)

55. M.G. Bellemare, Y. Naddaf, J. Veness, H.M. Bowling, The arcade learning environment: an evaluation platform for general agents. J. Artif. Intell. Res. **47**, 253–279 (2013)
56. M.G. Bellemare, J. Veness, M.H. Bowling, Investigating contingency awareness using Atari 2600 games, in *Twenty-Sixth AAAI Conference on Artificial Intelligence* (2012)
57. M. Hausknecht, J. Lehman, R. Miikkulainen, P. Stone, A neuroevolution approach to general atari game playing. IEEE Trans. Comput. Intell. AI Games **6**(4), 355–366 (2014)
58. J. Schulman, S. Levine, P. Moritz, M. Jordan, P. Abbeel, Trust region policy optimization, in *Proceedings of the 32nd International Conference on International Conference on Machine Learning* (2015)

Medical Image Segmentation Using Deep Neural Networks with Pre-trained Encoders

Alexandr A. Kalinin, Vladimir I. Iglovikov, Alexander Rakhlin and Alexey A. Shvets

Abstract With the growth of popularity of deep neural networks for image analysis, segmentation is the most common subject of studies applying deep learning to medical imaging and establishing state-of-the-art performance results in many applications. However, it still remains a challenging problem, for which performance improvements can potentially benefit diagnosis and other clinical practice outcomes. In this chapter, we consider two applications of multiple deep convolutional neural networks to medical image segmentation. First, we describe angiodysplasia lesion segmentation from wireless capsule endoscopy videos. Angiodysplasia is the most common vascular lesion of the gastrointestinal tract in the general population and is important to detect as it may indicate the possibility of gastrointestinal bleeding and/or anemia. As a baseline, we consider the U-Net model and then we demonstrate further performance improvements by using different deep architectures with ImageNet pre-trained encoders. In the second example, we apply these models to semantic segmentation of robotic instruments in surgical videos. Segmentation of instruments in the vicinity of surgical scenes is a challenging problem that is important for intraoperative guidance that can help the decision-making process. We achieve highly competitive performance for binary as well as for multi-class instrument segmentation. In both applications, we demonstrate that networks that employ ImageNet pre-trained encoders consistently outperform the U-Net architecture trained from scratch.

A. A. Kalinin (✉)
University of Michigan, Ann Arbor, MI 48109, USA
e-mail: akalinin@umich.edu

V. I. Iglovikov
ODS.ai, San Francisco, CA 94107, USA
e-mail: iglovikov@gmail.com

A. Rakhlin
Neuromation OU, 10111 Tallinn, Estonia
e-mail: rakhlin@neuromation.io

A. A. Shvets
Massachusetts Institute of Technology, Cambridge, MA 02142, USA
e-mail: shvets@mit.edu

M. A. Wani et al. (eds.), *Deep Learning Applications*,
Advances in Intelligent Systems and Computing 1098,
https://doi.org/10.1007/978-981-15-1816-4_3

1 Introduction

Segmentation of medical images is an important machine vision task that often has to be performed to enable computer-aided diagnostics and other downstream analyses. Recently, deep learning-based approaches demonstrated performance improvements over conventional machine learning methods for many problems in biomedical image analysis [1–3]. Specifically, deep convolutional neural networks have proven to be achieve state-of-the-art results for a broad range of medical image analysis tasks, such as breast cancer histology image analysis [4, 5], bone disease prediction [6], and age assessment [7]. Segmentation has been the most common subject of studies applying deep learning to medical imaging [3]. While these applications have demonstrated segmentation performance improvements, there is a need for further developments that potentially can significantly benefit clinical practice.

U-Net [8], introduced in 2015 and built upon the fully convolutional network [9], is arguably the most widely used deep neural network architecture for biomedical image segmentation, with over 7,000 citations to the original article according to Google Scholar as of mid-2019, making it the most cited publication in proceedings of the International Conference on Medical Image Computing and Computer Assisted Intervention (MICCAI). In general, U-Net architecture consists of the contracting path to capture context and the symmetrically expanding path that enables precise localization. The contracting path follows the typical architecture of a convolutional network with alternating convolution and pooling operations and progressively down-samples feature maps, increasing the number of feature maps per layer at the same time. Every step in the expansive path consists of an upsampling of the feature map followed by a convolution. Hence, the expansive branch increases the resolution of the output. In order to localize, upsampled features, the expansive path combines them with high-resolution features from the contracting path via skip-connections [8]. The output of the model is a pixel-by-pixel mask that shows the class of each pixel.

Typically, U-Net is trained from scratch starting with randomly initialized model weights. For the analysis of natural images, transfer learning has become a standard practice, when networks trained for classification of the ImageNet dataset [10] are employed as a method for model initialization in order to then reuse learnt weights for other tasks. It has been demonstrated that the use of ImageNet pre-training can improve model performance for classification of medical images, such as radiographs, despite the differences in pre-training and target image properties [11]. Iglovikov and Shvets [12] have shown that using networks trained on ImageNet as an encoder part of the U-Net architecture improved performance for satellite image segmentation.

In this chapter, we review two applications of deep neural networks with pre-trained encoders to medical image segmentation. Specifically, we describe angiodysplasia lesion segmentation from wireless capsule endoscopy videos [13] and semantic segmentation of robotic instrument from surgical videos [13]. We show that use

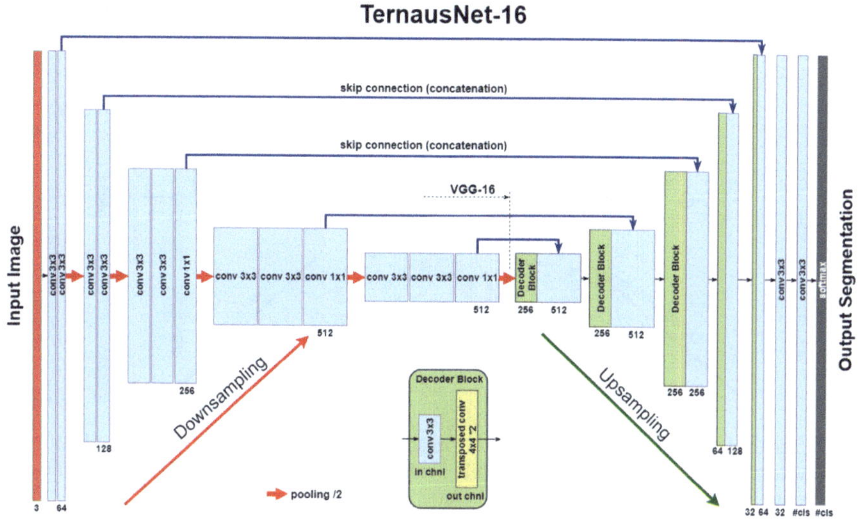

Fig. 1 Segmentation networks based on encoder–decoder architecture of U-Net family. TernausNet uses pre-trained VGG16 network as an encoder. Each box corresponds to a multi-channel feature map. The number of channels is pointed below the box. The height of the box represents a feature map resolution. The blue arrows denote skip-connections where information is transmitted from the encoder to the decoder

of ImageNet pre-trained networks as encoders consistently improves segmentation performance over the vanilla U-Net architecture trained from scratch.

2 Network Architectures and Training

In this chapter we consider four different deep architectures for segmentation: U-Net [8], two modifications of TernausNet [12], and LinkNet-34, based on the LinkNet model [14]. We use slightly modified version of the original U-Net model that previously proved itself successful in various segmentation problems with limited amounts of data, for example, see [7, 15].

As an improvement over the standard U-Net architecture, we use networks with similar general structure that employ different pre-trained encoders. TernausNet [12] is a U-Net-like architecture that uses relatively simple pre-trained VGG11 or VGG16 [16] networks as an encoder (see Fig. 1). VGG11 consists of seven convolutional layers, each followed by a ReLU activation function, and five max polling operations, each reducing feature map by 2. All convolutional layers have 3×3 kernels. TernausNet-16 has a similar structure and uses VGG16 network as an encoder (see Fig. 1).

In contrast, LinkNet-34 uses a pre-trained ResNet-34 [17] encoder, see Fig. 2. The encoder starts with the initial block that performs convolution with a kernel of size

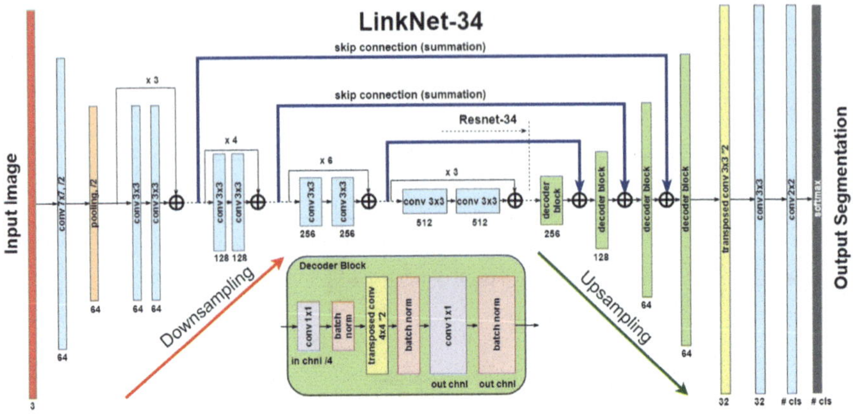

Fig. 2 LinkNet-34 uses pre-trained ResNet-34 as an encoder. It differs from TernausNet by adding skip-connections to the upsampling path, while TernausNet concatenates downsampled layers with the upsampling path (same as the original U-Net)

7×7 and stride 2. This block is followed by max-pooling with stride 2. The later portion of the network consists of repetitive residual blocks. In every residual block, the first convolution operation is implemented with stride 2 to provide downsampling, while the rest convolution operations use stride 1. In addition, the decoder of the network consists of several decoder blocks that are connected with the corresponding encoder block. As for TernausNets, the transmitted block from the encoder is concatenated to the corresponding decoder block. Each decoder block includes 1×1 convolution operation that reduces the number of filters by 4, followed by batch normalization and transposed convolution to upsample the feature map.

We use Jaccard index (Intersection Over Union) as the evaluation metric. It can be interpreted as a similarity measure between a finite number of sets. For two sets A and B, it can be defined as following:

$$J(A, B) = \frac{|A \cap B|}{|A \cup B|} = \frac{|A \cap B|}{|A| + |B| - |A \cap B|} \tag{1}$$

Since an image consists of pixels, the last expression can be adapted for discrete objects in the following way [12, 15]:

$$J = \frac{1}{n} \sum_{i=1}^{n} \left(\frac{y_i \hat{y}_i}{y_i + \hat{y}_i - y_i \hat{y}_i} \right), \tag{2}$$

where y_i and $\hat{y}_i$ are a binary value (label) and a predicted probability for the pixel i, correspondingly.

Since image segmentation task can also be considered as a pixel classification problem, we additionally use common classification loss functions, denoted as H.

For a binary segmentation problem H is binary cross entropy, while for a multi-class segmentation problem H is categorical cross entropy.

The final expression for the generalized loss function is obtained by combining (2) and H as following:

$$L = H - \log J \tag{3}$$

By minimizing this loss function, we simultaneously maximize probabilities for right pixels to be predicted and maximize the intersection J between masks and corresponding predictions. We refer reader to [15] for further details. Each model is trained with Adam optimizer [18] for 10 epochs with learning rate 0.001, and then for another 5 epochs with the learning rate 0.0001.

As an output of a model, we obtain an image, in which each pixel value corresponds to a probability of belonging to the area of interest or a class. The size of the output image matches the input image size. For two applications we used different pre- and post-processing procedures, as described below in their corresponding sections.

3 Angiodysplasia Lesion Segmentation in Wireless Capsule Endoscopy Videos

3.1 Background

Angiodysplasia (AD) is the most common vascular lesion of the gastrointestinal (GI) tract in the general population [19]. This condition may be asymptomatic, or it may cause gastrointestinal bleeding or and anemia [20]. Wireless capsule endoscopy (WCE) is the preferred first-line investigation for the small bowel in the context of GI bleeding as it is safe, acceptable and has significantly higher or at least equivalent yield for lesions when compared with alternative methods [21, 22]. However, only 69% of angiodysplasias are detected by gastroenterologist experts during the reading of WCE videos, and blood indicator software (supplied by a WCE provider), in the presence of angiodysplasias, presents sensitivity and specificity values of only 41% and 67%, respectively [23]. Therefore, there is a compelling need to improve accuracy of AD detection and localization for the potential use in clinical practice.

There is a number of computer vision-based methods developed for the video capsule endoscopy analysis [24], including rule-based and conventional machine learning algorithms that are applied to extracted color, texture, and other features [25–27].

In this chapter, we review an application of deep convolutional neural networks for angiodysplasia lesions segmentation from video capsule endoscopy [13]. First, we describe the approach that utilizes the original U-Net architecture. This method was used to produce a submission to the MICCAI 2017 Endoscopic Vision SubChallenge: Angiodysplasia detection and localization [23] that placed first, winning the competition. Then, we review further improvements over this solution by utilizing deep convolutional neural networks with pre-trained encoders: TernausNet [12] and LinkNet-34.

3.2 Dataset Description and Preprocessing

A wireless capsule endoscope, is a disposable plastic capsule that weights 3.7 g and measures 11 mm in diameter and 26 mm in length. Image features include a 140° field of view, 1:8 magnification, 1–30 mm depth of view, and a minimum size of detection of about 0.1 mm. The capsule is passively propelled through the intestine by peristalsis while transmitting color images. Last generation of this device is able to acquire more than 60,000 images with a resolution of approximately 520×520 pixels [28].

The dataset consists of 1200 color images obtained with WCE. The images are in the 24-bit PNG format, with the 576×576 pixel resolution. The dataset is split into two equal parts, 600 images for training and 600 for evaluation. Each subset is composed of 300 images with apparent AD and 300 without any pathology. The training subset is annotated by human expert and contains 300 binary masks in JPEG format of the same 576×576 pixel resolution. White pixels in the masks correspond to lesion localization. Several examples from the training set are given in Fig. 3, where the first row corresponds to images without pathology, the second one to images with several AD lesions in every image, and the last row contains masks that correspond to the pathology images from the second row. In the dataset each image contains up to 6 lesions and their distribution is shown in Fig. 4 (left). As shown, the most images contain only 1 lesion. In addition, Fig. 4 (right) shows distribution of AD lesion areas that reach the maximum of approximately 12,000 pixels with the median value of 1,648 pixels.

Images are cropped from 576×576 to 512×512 pixels to remove the canvas and text annotations. Then we rescale the data from $[0 \ldots 255]$ to $[0 \ldots 1]$ and standardize it following the ImageNet scheme [12]. For training and cross-validation, we only use 299 images annotated with binary masks that contain pathology. With those, we randomly split the dataset into fivefolds of 60, 60, 60, 60, and 59 images. In order to improve model generalization during training, random affine transformations and color augmentations in HSV space are applied.

Following the segmentation step, we perform postprocessing in order to find the coordinates of angiodysplasia lesions in the image. In the postprocessing step we use the OpenCV implementation of a connected component labeling function: `connectedComponentsWithStats` [29]. This function returns the number of connected components, their sizes (areas), and centroid coordinates of the corresponding connected component. In our detector we use another threshold to neglect all clusters with the size smaller than 300 pixels. Therefore, in order to establish the presence of the lesions, the number of found components should be higher than 0, otherwise the image corresponds to a normal condition. Then, for localization of angiodysplasia lesions we return centroid coordinates of all connected components.

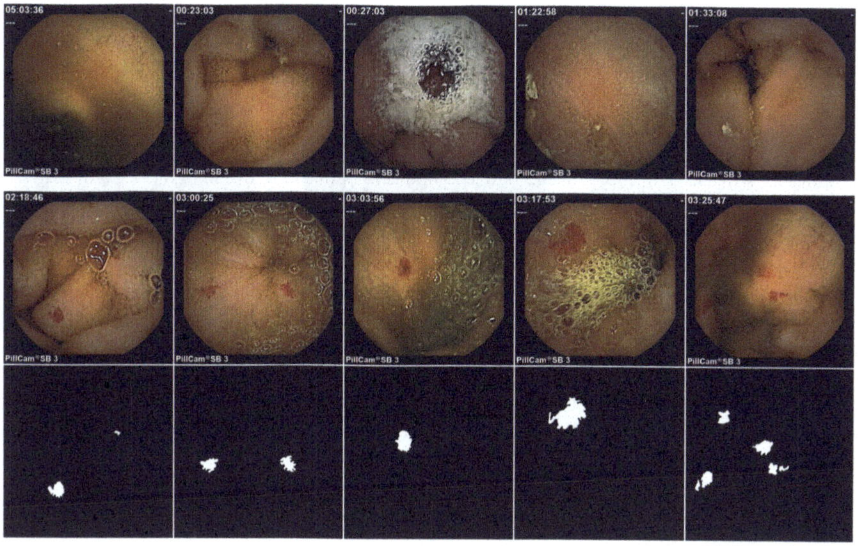

Fig. 3 Sample images from the training set for Angiodysplasia detection and localization challenge [23]. The upper row corresponds to normal images. In the middle row the images contain angiodysplasia area represented as red spots. The down row contains masks for angiodysplasia from the middle row

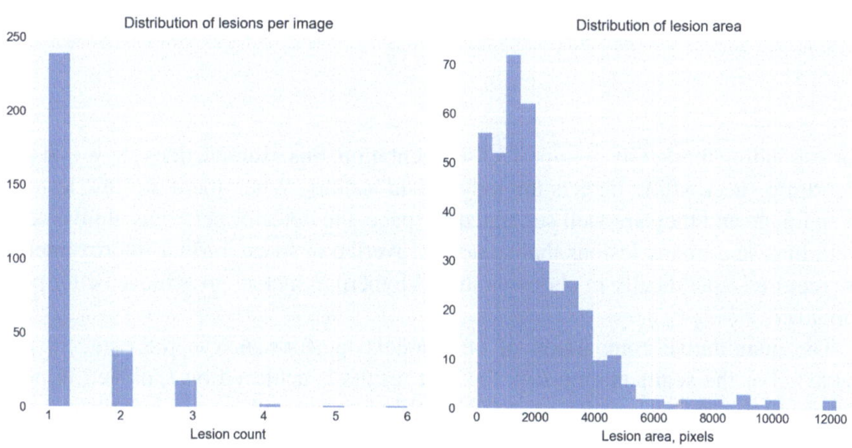

Fig. 4 Distribution of angiodysplasia lesions per image (left figure) and distribution of lesions area (right figure) in the data set

3.3 Results

To test our prediction and compare it with known mask we performed calculations on the image taken from the validation set. The exemplar result of the prediction is shown in Fig. 5. For a visual comparison we also provide the original image and its

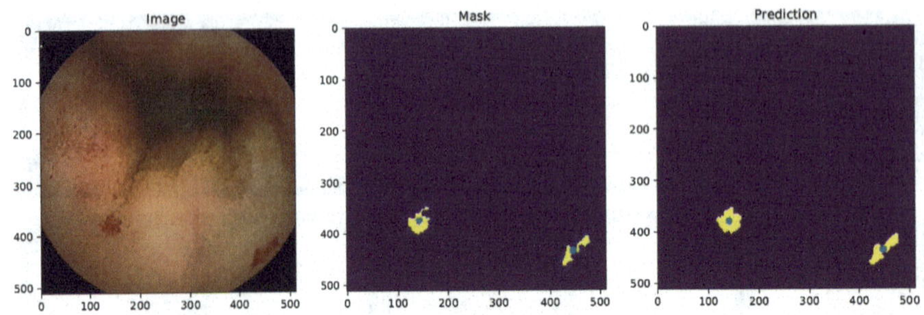

Fig. 5 The prediction of our detector on the validation data image. Here, the first picture corresponds to original image, the second one to the training mask, the last one to the predicted mask. Green dots inside the clusters correspond to the centroid coordinates that define a localization of the appropriate angiodysplasia. For example, the real values for centroid coordinates are p^1_{mask} = (376, 144), p^1_{pred} = (380, 143) for the first cluster and p^2_{mask} = (437, 445), p^2_{pred} = (437, 447) for the second one

Table 1 Segmentation results. Intersection over Union (IoU) and Dice coefficient (Dice) are in % and inference time (Time) is in *ms*

Model	IOU	Dice	Time
U-Net	73.18	83.06	30
TernausNet-11	74.94	84.43	51
TernausNet-16	73.83	83.05	60
LinkNet-34	**75.35**	**84.98**	**21**

corresponding mask. Given imperfect segmentation, this example does show that the algorithm successfully detects angiodysplasia lesions. When there are few lesions in an image and they are well separated in space, the detector performs almost very well. In case of many lesions that somehow overlap in space, further improvements are required, specifically in choosing model hyperparameters, to achieve better performance.

The quantitative comparison of our models' performance is presented in the Table 1. For the segmentation task the best results is achieved by LinkNet-34 providing $IoU = 0.754$ and $Dice = 0.831$. When compared by the inference time, LinkNet-34 is also the fastest model due to the light encoder. In the segmentation task this network takes around 20 ms for 512×512 pixel image and more than three times as fast as TernausNets. The inference speed is important for timely processing of large amounts of frames that can be obtained with a WCE that defines how quickly corresponding downstream diagnosis can be made by a gastroenterologist. In these experiments, the inference time was measured using one NVIDIA GTX 1080Ti GPU.

The corresponding code was made publicly available under MIT licence at https://github.com/ternaus/angiodysplasia-segmentation.

4 Robotic Instrument Segmentation in Surgical Videos

4.1 Background

Information in a surgical console of a robot-assisted surgical system includes valuable details for intraoperative guidance that can help the decision-making process. This information is usually represented as 2D images or videos that contain surgical instruments and patient tissues. Understanding these data is a complex problem that involves the tracking and pose estimation for surgical instruments in the vicinity of surgical scenes. A critical component of this process is semantic segmentation of the instruments in the surgical console. Semantic segmentation of robotic instruments is a difficult task by the virtue of light changes such as shadows and specular reflections, visual occlusions such as blood and camera lens fogging, and due to the complex and dynamic nature of background tissues [30]. Segmentation masks can be used to provide a reliable input to instrument tracking systems. Therefore, there is a compelling need for the development of accurate and robust computer vision methods for semantic segmentation of surgical instruments from operational images and video.

There is a number of vision-based methods developed for the robotic instrument detection and tracking [30]. Instrument-background segmentation can be treated as a binary or instance segmentation problem for which classical machine learning algorithms have been applied using color and/or texture features [31, 32]. Later applications addressed this problem as semantic segmentation, aiming to distinguish between different instruments or their parts [33, 34]. Previous deep learning-based applications to robotic instrument segmentation have demonstrated competitive performance in binary segmentation [35, 36] and promising results in multi-class segmentation [37].

In this chapter, we review an application of deep convolutional neural networks for robotic instrument semantic segmentation from surgical videos [13]. First, we consider the approach that utilizes the original U-Net architecture. This method was used to produce a submission to the MICCAI 2017 Endoscopic Vision SubChallenge: Robotic Instrument Segmentation [38]. This submission placed first in binary and multi-class instrument segmentation and second in instrument parts segmentation sub-tasks placed first, winning the competition. Then, we provide the details about further improvements over this solution by utilizing deep convolutional neural networks with pre-trained encoders: TernausNet [12] and LinkNet-34.

4.2 Dataset Description and Preprocessing

The training dataset consists of 8×225-frame sequences of high-resolution stereo camera images acquired from a da Vinci Xi surgical system during several different porcine procedures [38]. Training sequences are provided with 2 Hz frame rate to avoid redundancy. Every video sequence consists of two stereo channels taken

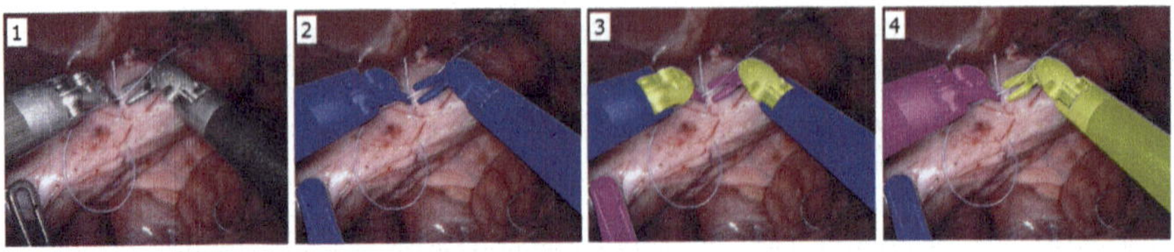

Fig. 6 A snapshot from a robotic surgical video that contains robotic instruments and patient tissues: **1** original video frame; **2** binary segmentation of robotic instruments shown in blue and tissue that serves as a background; **3** multi-class segmentation of robotic instruments where each class corresponds to a different part of the robotic instrument (3 classes: rigid shaft, articulated wrist and claspers); and **4** multi-class segmentation of robotic instruments where each class corresponds to a different robotic instrument (7 classes)

from left and right cameras and has a 1920 × 1080 pixel resolution in the RGB format. To remove black canvas and extract original 1280 × 1024 camera images from the frames, an image has to be cropped starting from the pixel at the (320, 28) position. Ground truth labels are provided for left frames only, therefore only left channel images are used for training. The articulated parts of the robotic surgical instruments, such as a rigid shaft, an articulated wrist and claspers have been hand-labeled in each frame. Ground truth labels are encoded with numerical values (10, 20, 30, 40, 0) and assigned to each part of an instrument or background. Furthermore, there are instrument type labels that categorize instruments in the following categories: left/right prograsp forceps, monopolar curved scissors, large needle driver, and a miscellaneous category for any other surgical instruments, see Fig. 6.

The test dataset consists of 8 × 75-frame sequences containing footage sampled immediately after each training sequence and 2 full 300-frame sequences, sampled at the same rate as the training set. Under the terms of the challenge, participants should exclude the corresponding training set when evaluating one of the 75-frame sequences.

As an output of a model, we obtained an image, in which each pixel value corresponds to a probability of belonging to the area of interest or a class. The size of the output image matches the input image size. For binary segmentation, we used 0.3 as a threshold value (chosen using validation dataset) to binarize pixel probabilities. All pixel values below the specified threshold were set to 0, while all values above the threshold were set to 255 to produce final prediction mask. For multi-class segmentation we used a similar procedure, but we set different integer numbers for each class, as was noted above.

4.3 Results

The qualitative comparison of our models both for a binary and multi-class segmentation is presented in Fig. 7 and Table 2. For the binary segmentation task the

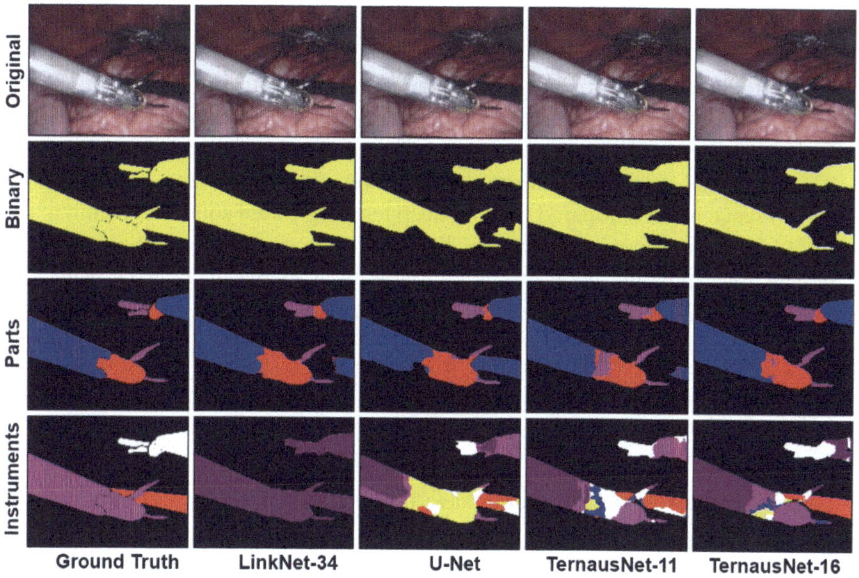

Fig. 7 Qualitative comparison between several neural network architectures implemented for a binary and multi-class segmentation

Table 2 Segmentation results per task. Intersection over Union (IoU) and Dice coefficient (Dice) are in % and inference time (Time) is in *ms*

	Binary segmentation			Parts segmentation			Instrument segmentation		
Model	IOU	Dice	Time	IOU	Dice	Time	IOU	Dice	Time
U-Net	75.44	84.37	93	48.41	60.75	106	15.80	23.59	**122**
TernausNet-11	81.14	88.07	142	62.23	74.25	157	**34.61**	**45.86**	173
TernausNet-16	**83.60**	**90.01**	184	**65.50**	**75.97**	202	33.78	44.95	275
LinkNet-34	82.36	88.87	**88**	34.55	41.26	**97**	22.47	24.71	177

best results is achieved by TernausNet-16 with $IoU = 0.836$ and $Dice = 0.901$. For multi-class segmentation of different parts of instruments, the best results are also by TernausNet-16 with $IoU = 0.655$ and $Dice = 0.760$. For the multi-class class instrument segmentation task the results look less optimistic. In this case the best model is TernausNet-11 that achieves $IoU = 0.346$ and $Dice = 0.459$ for segmentation on 7 classes. Lower performance can be explained by the relatively small dataset size. There are 7 classes and several classes appear just few times in the training dataset. The results look different from the previous application, where LinkNet-34 was the best in binary segmentation, possibly due to the same limitation. Thus, the results suggest that the performance can be improved by increasing the dataset size.

When compared by the inference time, LinkNet-34 is still the fastest model due to the light encoder. In the case of a binary segmentation task this network takes around $90\,ms$ for 1280×1024 pixel image and more than twice as fast as TernausNet. Same as in the previous application, the inference time was measured using one NVIDIA GTX 1080Ti GPU.

This suggested approach demonstrated improvement over the state-of-the-art level of performance when compared to other deep learning-based solutions within to the MICCAI 2017 Endoscopic Vision SubChallenge: Robotic Instrument Segmentation [38]. The corresponding code was made publicly available under MIT licence at https://github.com/ternaus/robot-surgery-segmentation.

5 Conclusions

Deep convolutional neural networks have become an approach of choice for segmentation of various biomedical images. U-Net architecture provides a very strong baseline performance in segmentation tasks even when the amount of available labeled data is limited. Borrowing the idea from transfer learning approaches to image classification, the use of pre-trained networks as encoders inside a U-Net-like network further improves model performance. We review two separate applications of such networks. First, models with encoders based on ImageNet pre-trained VGG-11, VGG-16, and ResNet-34 networks outperform U-Net trained from scratch for angiodysplasia lesion segmentation from wireless capsule endoscopy videos. In this instance, LinkNet-34 is the best performing model that also provides the fastest inference speed. In the second application, the same models also outperform vanilla U-Net for the semantic segmentation of robotic instruments in surgical videos. Here, the best performing model was TernausNet-16 with pre-trained VGG-16 encoder. These results suggest that deeper encoders demonstrate better results given more training data for segmentation, since weights of pre-trained layers are being fine-tuned as well. Further possible improvements may involve the use of application-specific image augmentations [39] and even deeper pre-trained encoders given enough labeled data.

References

1. H. Greenspan, B. Van Ginneken, R.M. Summers, Guest editorial deep learning in medical imaging: Overview and future promise of an exciting new technique. IEEE Trans. Med. Imaging **35**(5), 1153–1159 (2016)
2. T. Ching, D.S. Himmelstein, B.K. Beaulieu-Jones, A.A. Kalinin, B.T. Do, G.P. Way, E. Ferrero, P.-M. Agapow, M. Zietz, M.M. Hoffman, W. Xie, G.L. Rosen, B.J. Lengerich, J. Israeli, J. Lanchantin, S. Woloszynek, A.E. Carpenter, A. Shrikumar, J. Xu, E.M. Cofer, C.A. Lavender, S.C. Turaga, A.M. Alexandari, Z. Lu, D.J. Harris, D. DeCaprio, Y. Qi, A. Kundaje, Y. Peng, L.K. Wiley, M.H.S. Segler, S.M. Boca, S.J. Swamidass, A. Huang, A. Gitter, C.S. Greene,

Opportunities and obstacles for deep learning in biology and medicine. J. R. Soc. Interface **15**(141) (2018)
3. G. Litjens, T. Kooi, B.E. Bejnordi, A.A.A. Setio, F. Ciompi, M. Ghafoorian, J.A. Van Der Laak, B. Van Ginneken, C.I. Sánchez, A survey on deep learning in medical image analysis. Med. Image Anal. **42**, 60–88 (2017)
4. A. Rakhlin, A. Shvets, V. Iglovikov, A.A. Kalinin, Deep convolutional neural networks for breast cancer histology image analysis, in *International Conference Image Analysis and Recognition* (Springer, 2018), pp. 737–744
5. A. Rakhlin, A.A. Shvets, A.A. Kalinin, A. Tiulpin, V.I. Iglovikov, S. Nikolenko, Breast tumor cellularity assessment using deep neural networks, in *2019 IEEE International Conference on Computer Vision Workshops (ICCVW)* (IEEE, 2019)
6. A. Tiulpin, J. Thevenot, E. Rahtu, P. Lehenkari, S. Saarakkala, Automatic knee osteoarthritis diagnosis from plain radiographs: a deep learning-based approach. Sci. Rep. **8**, 1727 (2018)
7. V.I. Iglovikov, A. Rakhlin, A.A. Kalinin, A.A. Shvets, Paediatric bone age assessment using deep convolutional neural networks, in *Deep Learning in Medical Image Analysis and Multimodal Learning for Clinical Decision Support* (Springer, 2018), pp. 300–308
8. O. Ronneberger, P. Fischer, T. Brox, U-net: convolutional networks for biomedical image segmentation, in *International Conference on Medical Image Computing and Computer-Assisted Intervention* (Springer, 2015), pp. 234–241
9. J. Long, E. Shelhamer, T. Darrell, Fully convolutional networks for semantic segmentation, in *Proceedings of the IEEE Conference on Computer Vision and Pattern Recognition* (2015), pp. 3431–3440
10. O. Russakovsky, J. Deng, H. Su, J. Krause, S. Satheesh, S. Ma, Z. Huang, A. Karpathy, A. Khosla, M. Bernstein et al., Imagenet large scale visual recognition challenge. Int. J. Comput. Vis. **115**(3), 211–252 (2015)
11. A. Rajkomar, S. Lingam, A.G. Taylor, M. Blum, J. Mongan, High-throughput classification of radiographs using deep convolutional neural networks. J. Digit. Imaging **30**(1), 95–101 (2017)
12. V. Iglovikov, A. Shvets, Ternausnet: U-net with vgg11 encoder pre-trained on imagenet for image segmentation (2018), arXiv:1801.05746
13. A.A. Shvets, A. Rakhlin, A.A. Kalinin, V.I. Iglovikov, Automatic instrument segmentation in robot-assisted surgery using deep learning, in *2018 17th IEEE International Conference on Machine Learning and Applications (ICMLA)* (IEEE, 2018), pp. 624–628
14. A. Chaurasia, E. Culurciello, Linknet: Exploiting encoder representations for efficient semantic segmentation (2017), arXiv:1707.03718
15. V. Iglovikov, S. Mushinskiy, V. Osin, Satellite imagery feature detection using deep convolutional neural network: a kaggle competition (2017), arXiv:1706.06169
16. K. Simonyan, A. Zisserman, Very deep convolutional networks for large-scale image recognition (2014), arXiv:1409.1556
17. K. He, X. Zhang, S. Ren, J. Sun, Deep residual learning for image recognition, in *Proceedings of the IEEE Conference on Computer Vision and Pattern Recognition* (2016), pp. 770–778
18. D.P. Kingma, J. Ba, Adam: a method for stochastic optimization (2014), arXiv:1412.6980
19. P.G. Foutch, D.K. Rex, D.A. Lieberman, Prevalence and natural history of colonic angiodysplasia among healthy asymptomatic people. Am. J. Gastroenterol. **90**(4) (1995)
20. J. Regula, E. Wronska, J. Pachlewski, Vascular lesions of the gastrointestinal tract. Best Pract. Res. Clin. Gastroenterol. **22**(2), 313–328 (2008)
21. S.L. Triester, J.A. Leighton, G.I. Leontiadis, D.E. Fleischer, A.K. Hara, R.I. Heigh, A.D. Shiff, V.K. Sharma, A meta-analysis of the yield of capsule endoscopy compared to other diagnostic modalities in patients with obscure gastrointestinal bleeding. Am. J. Gastroenterol. **100**(11), 2407 (2005)
22. R. Marmo, G. Rotondano, R. Piscopo, M. Bianco, L. Cipolletta, Meta-analysis: capsule enteroscopy vs. conventional modalities in diagnosis of small bowel diseases. Aliment. Pharmacol. Ther. **22**(7), 595–604 (2005)
23. MICCAI 2017 Endoscopic Vision Challenge: Angiodysplasia Detection and Localization, https://endovissub2017-giana.grand-challenge.org/angiodysplasia-etisdb/

24. D.K. Iakovidis, A. Koulaouzidis, Software for enhanced video capsule endoscopy: challenges for essential progress. Nat. Rev. Gastroenterol. Hepatol. **12**(3), 172 (2015)
25. M. Mackiewicz, J. Berens, M. Fisher, Wireless capsule endoscopy color video segmentation. IEEE Trans. Med. Imaging **27**(12), 1769–1781 (2008)
26. A. Karargyris, N. Bourbakis, Wireless capsule endoscopy and endoscopic imaging: a survey on various methodologies presented. IEEE Eng. Med. Biol. Mag. **29**(1), 72–83 (2010)
27. P. Szczypiński, A. Klepaczko, M. Pazurek, P. Daniel, Texture and color based image segmentation and pathology detection in capsule endoscopy videos. Comput. Methods Programs Biomed. **113**(1), 396–411, (2014), http://www.sciencedirect.com/science/article/pii/S0169260712002192
28. D.S. Mishkin, R. Chuttani, J. Croffie, J. DiSario, J. Liu, R. Shah, L. Somogyi, W. Tierney, L.M.W.K. Song, B.T. Petersen, Asge technology status evaluation report: wireless capsule endoscopy. Gastrointest. Endosc. **63**(4), 539–545 (2006)
29. G. Bradski, The OpenCV Library, in *Dr. Dobb's Journal of Software Tools* (2000)
30. B. Münzer, K. Schoeffmann, L. Böszörmenyi, Content-based processing and analysis of endoscopic images and videos: a survey. Multimed. Tools Appl. **77**(1), 1323–1362 (2018)
31. S. Speidel, M. Delles, C. Gutt, R. Dillmann, Tracking of instruments in minimally invasive surgery for surgical skill analysis, in *Medical Imaging and Augmented Reality* (Springer, Berlin, 2006), pp. 148–155
32. C. Doignon, F. Nageotte, M. De Mathelin, Segmentation and guidance of multiple rigid objects for intra-operative endoscopic vision, in *Dynamical Vision*. (Springer, Berlin, 2007), pp. 314–327
33. Z. Pezzementi, S. Voros, G.D. Hager, Articulated object tracking by rendering consistent appearance parts, in *IEEE International Conference on Robotics and Automation, 2009. ICRA'09.* (IEEE, 2009), pp. 3940–3947
34. D. Bouget, R. Benenson, M. Omran, L. Riffaud, B. Schiele, P. Jannin, Detecting surgical tools by modelling local appearance and global shape. IEEE Trans. Med. Imaging **34**(12), 2603–2617 (2015)
35. L.C. García-Peraza-Herrera, W. Li, L. Fidon, C. Gruijthuijsen, A. Devreker, G. Attilakos, J. Deprest, E.B.V. Poorten, D. Stoyanov, T. Vercauteren, S. Ourselin, Toolnet: Holistically-nested real-time segmentation of robotic surgical tools, in *Proceedings of the 2017 IEEE/RSJ International Conference on Intelligent Robots and Systems* (2017), pp. 5717–5722
36. M. Attia, M. Hossny, S. Nahavandi, H. Asadi, Surgical tool segmentation using a hybrid deep cnn-rnn auto encoder-decoder, in *2017 IEEE International Conference on Systems, Man, and Cybernetics (SMC)* (2017), pp. 3373–3378
37. D. Pakhomov, V. Premachandran, M. Allan, M. Azizian, N. Navab, Deep residual learning for instrument segmentation in robotic surgery (2017), arXiv:1703.08580
38. M. Allan, A. Shvets, T. Kurmann, Z. Zhang, R. Duggal, Y.-H. Su, N. Rieke, I. Laina, N. Kalavakonda, S. Bodenstedt, et al., 2017 robotic instrument segmentation challenge (2019), arXiv:1902.06426
39. A. Buslaev, A. Parinov, E. Khvedchenya, V.I. Iglovikov, A.A. Kalinin, Albumentations: fast and flexible image augmentations (2018), arXiv:1809.06839

Ensemble of 3D Densely Connected Convolutional Network for Diagnosis of Mild Cognitive Impairment and Alzheimer's Disease

Shuqiang Wang, Hongfei Wang, Albert C. Cheung, Yanyan Shen and Min Gan

Abstract Automatic diagnosis of Alzheimer's disease (AD) and mild cognitive impairment (MCI) from 3D brain magnetic resonance (MR) images play an important role in the early treatment of dementia disease. Deep learning architectures can extract potential features of dementia disease and capture brain anatomical changes from MRI scans. Given the high dimension and complex features of the 3D medical images, computer-aided diagnosis is still confronted with challenges. Firstly, compared with the number of learnable parameters, the number of training samples is very limited, which can cause overfitting problems. Secondly, the deepening of the network layer makes gradient information gradually weaken and even disappears in the process of transmission, resulting in mode collapse. This chapter proposed an ensemble of 3D densely connected convolutional networks for AD and MCI diagnosis from 3D MRIs. Dense connections were introduced to maximize the information flow, where each layer connects with all subsequent layers directly. Bottleneck layers and transition layers are also employed to reduce parameters and lead to more compact models. Then the probability-based fusion method was employed to combine 3D-DenseNets with different architectures. Extensive experiments were conducted to analyze the performance of 3D-DenseNet with different hyperparameters and architectures. Superior performance of the proposed model was demonstrated on ADNI dataset.

S. Wang · H. Wang · Y. Shen
Shenzhen Institutes of Advanced Technology, Chinese Academy of Sciences, 1068 Xueyuan Avenue, Shenzhen University Town, Shenzhen, China
e-mail: sq.wang@siat.ac.cn

A. C. Cheung
Hong Kong University of Science and Technology, Hong Kong SAR, China

M. Gan (✉)
College of Mathematics and Computer Science, Fuzhou University, Fuzhou 350116, China
e-mail: aganmin@aliyun.com

M. A. Wani et al. (eds.), *Deep Learning Applications*, Advances in Intelligent Systems and Computing 1098,
https://doi.org/10.1007/978-981-15-1816-4_4

1 Introduction

Alzheimer's disease is a common progressive neurodegenerative disease, which has been listed as the fourth biggest killer threatening the life and health of the elderly in developed countries. AD is caused by the damage and destruction of nerve cells in brain regions which are related to memory, and its most common symptoms are memory loss and cognitive decline [1]. Mild cognitive impairment (MCI) has been known as an intermediate transition between normal elderly and AD [2]. In a recent research, 32% of patients with MCI worsened Alzheimer's disease within 5 years. Therefore, the early diagnosis and intervention of MCI and AD are of great significance in controlling the development of the disease.

It is difficult to achieve accurate diagnosis of AD and MCI through cognitive tests and clinical symptoms, but the structural and functional information of the brain can be obtained by means of neuroimaging such as magnetic resonance imaging (MRI), computed tomography (CT) and positron emission tomography (PET), and the diagnosis is quite reliable. In the clinic, medical image interpretation mostly relies on human experts such as radiologists and physicians. Because of wide variations in pathology and the potential fatigue of doctors, researchers and medical experts begun to benefit from computer-aided diagnosis (CAD). Over the past decades, neuroimaging data have been increasingly used to characterize AD and MCI by means of machine learning (ML) methods, offering promising tools for individualized diagnosis and prognosis. Numerous studies have been proposed to use predefined features (including regional and voxel-based measurements) from image preprocessing pipelines followed by different types of classifiers, such as support vector machines (SVM) or random forests. Recently, deep learning (DL), as a newly emerging modeling methodology, has made a big leap in the domain of medical imaging [3–5]. Particularly, deep convolutional neural networks (CNNs) have been proved to be excellent in the automatic diagnosis of cognitive disease from brain MR images, in that deep CNN has proven to be a powerful method for learning abstract features from raw data. Compared with 2D convolutions on slices, 3D convolutions on a whole MRI can capture potential 3D structural information which may be essential for discrimination. Due to the complex structure of 3D MRI and its high-dimensional features, the 3D-CNNs would be designed deeper to model high-level abstractions of the data. However, the performance of 3D-CNNs is very limited when the gradient information passes through many layers, because the gradient information may vanish during the forward and backward propagation. In addition, a tremendous amount of parameters, which refers to the weights of convolution kernels, cannot be optimized entirely through the limited training set.

In this chapter, we proposed an ensemble of 3D densely connected convolutional networks for AD and MCI diagnosis. Dense connections were introduced to improve the feature utilization, then the network could be deeper due to less feature increment in each layer and fewer parameters. What is more, the probability-based ensemble approach can decrease the misrecognition risk of selecting a single classifier and is becoming popular for medical image analysis.

2 Related Work

2.1 Deep Learning for Computer-Aided Diagnosis

The rapid development of deep learning research has attracted the attention of medical image analysis experts. CAD systems have achieved amazing results in the field of medical image analysis by virtue of their excellent automatic feature learning and nonlinear modeling capabilities. CADs extract and model features automatically from medical images, then provide an objective opinion regarding assessment of a disease. The major applications of CAD include discrimination of malign to benign lesions and the identification of certain diseases from one or more images. Cheng et al. [6] used an SAE with a denoising technique (SDAE) to recognize breast ultrasound lesions and lung CT nodules. Plis et al. [7] developed a DBN analysis method for MR images and validated the feasibility of the application by investigating whether a building block of deep generative models was competitive with independent component analysis, the most widely used method for functional MRI (fMRI) analysis. Gheus et al. [8] proposed a sparse adaptive neural network based on edge learning for aortic valve detection in echocardiographic images, which alleviated the high complexity of 3D image data. Shen et al. [9] proposed a hierarchical learning model with a multi-scale conventional network to capture different sizes of lung nodules. In this CNN architecture, three CNNs that took nodule patches from various scales as input were assembled in parallel, which improves recognition accuracy greatly. Wang et al. [10] proposed an automated skeletal maturity recognition system that takes a single hand radiograph as input and finally output the bone age prediction.

2.2 Automatic Recognition of Alzheimer's Disease and Mild Cognitive Impairment

Precomputed medical descriptors together with statistical and conventional machine learning methods have been widely used to automatic diagnosis of AD. Risacher et al. [11] calculated the hippocampal volumes gray matter (GM) density, and cortical thickness values from segmented regions of interest (ROI). Then voxel-based morphometry (VBM) method was used for MRI analysis and AD classification. Cai et al. [12] designed 3D pathology-centric masks to extract the cerebral metabolic rate of glucose consumption (CMRGlc), and a content-based retrieval method was proposed for 3D neurological images. Liu et al. [13] extracted 83 ROIs from 3D brain MRI and PET scans, and proposed a Multifold Bayesian Kernelization (MBK) based on a Bayesian framework to diagnose AD. Zhang et al. [14] extracted 93 ROIs from the MRI and PET scans using standardized templates, which were designed by human experts on the basis of their knowledge about the target domains, and the multimodal features were combined through a multi-kernel support vector machine (SVM). Zhang et al. [15] used k-means clustering to build a low-level features dictionary involving lesion metabolism index, mean index, and Fisher index. Then, Probabilistic Latent Semantic Analysis (PLSA) and Canonical Correlation Analysis (CCA) were used to combine the features and capture the latent associations.

Compared with other machine learning techniques mentioned above, deep learning has witnessed significant advances. Liu et al. [16] trained a deep neural network contained autoencoders to combine multimodal features which were extracted from 83 ROIs of PET and MRI scans. Li et al. [17] achieved multimodal fusion of PET and MRI features through a restricted Boltzmann machine (RBM) and improved the classification accuracies by designing a multitask deep learning network with dropout. ROI-based methods can significantly extract representative features and partly reduce the feature dimension, but the ROIs are too empirical to capture the larvaceous features entirely which are associated with AD diagnose. Convolutional neural networks (CNNs) have been widely used in pattern recognition and present an outstanding performance on AD classification through medical images. Billones et al. [18] selected 20 successive slices from MRI, under the hypothesis that the slices cover the significant areas for dementia detection. And each serial number of the 2D slices was used to train a 2D-CNN, respectively modified from the VGGNet. 3D-CNN can capture more complete spatial features through its space association ability. Hosseini-Asl et al. [19] proposed a 3D convolutional neural network which combines a 3D convolutional autoencoder pretrained with registered images. Payan et al. [20] built a learning algorithm by combining 3D convolutions and sparse autoencoders, and used it on a whole MRI. Cheng et al. [21] extracted a number of 3D patches from the whole MRI and a patch transformed into features by 3D-CNN. Finally, multiple 3D-CNNs were used to combine the features and demonstrated the effectiveness of AD classification.

3 Methods

3.1 Data Acquisition and Preprocessing

In this work, we obtained neuroimaging data from the Alzheimer's Disease Neuroimaging Initiative (ADNI) database [22]. The study involves more than 1,000 participants including people with MCI, patients with diagnosed AD and normal contrasts. Most of the participants were collected repeatedly for two to six times, and the interval between neighbor scans was more than a year. Time sequence scans provide researchers with a novel discovery of AD progression.

As shown in Fig. 1, a total of 833 T1-weighted MRIs were employed, which were collected from 624 participants, including both male and female, and their ages ranging from 70 to 90. Since a given participant's brain structure makes a difference after a period of time, we selected two scans with the longest interval of one participant as different subjects, as long as the interval is more than 3 years. And when 10-fold cross-validation was employed to evaluate our models, the subjects selected from the same participant were bound up and placed in the same subgroup so that they were forbidden to appear both in training and testing datasets.

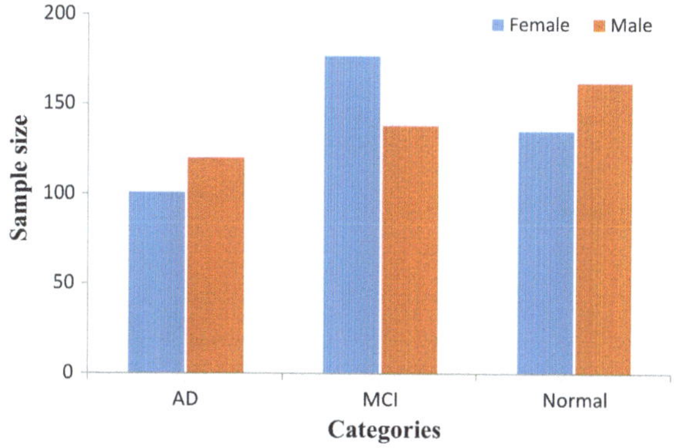

Fig. 1 Distribution of the subjects

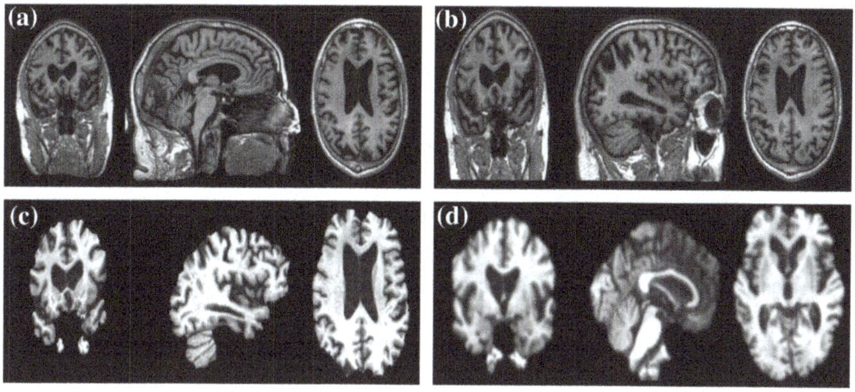

Fig. 2 Images after each prepressing step: **a** Original image; **b** Image after removal of redundant tissues; **c** Image after brain extraction; **d** Image aligned to the MIN152 template

The selected MRIs were considered the best in the quality ratings and have undergone grad-warping, intensity correction, and have been scaled for gradient drift using the phantom data. Brain Extraction Tool (FSL-BET) [23] was used to strip non-brain tissue from an image of the whole head, which can reduce classification errors caused by redundant information. In order to make anatomical points of different images aligned, images were registered to the standardized template using FSL FLIRT [24]. The dimension of each image is 91 × 109 × 91 in Neuroimaging Informatics Technology Initiative (NIfTI) file format. Preprocessed steps are shown in Fig. 2.

3.2 *Proposed Method*

Consider a traditional network comprises l layers, we denote x_l as the output of the lth layer, and each layer implements a nonlinear transformation $H_l(\cdot)$, where l indexes

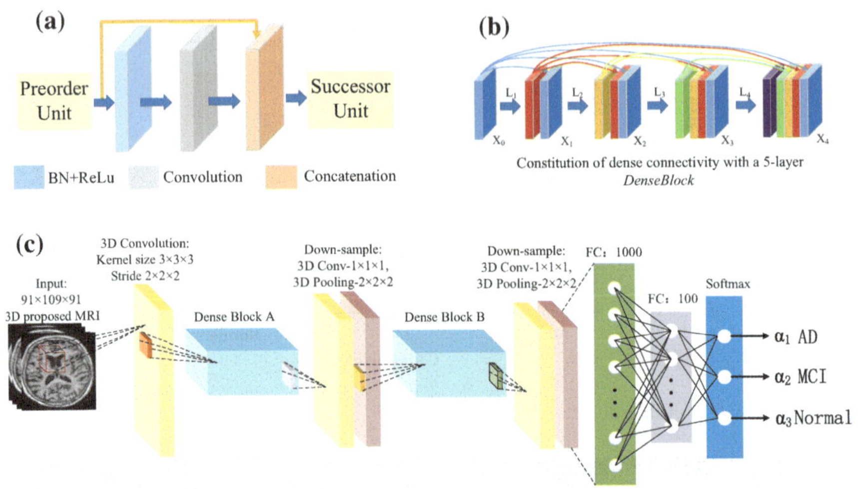

Fig. 3 **a** A composite dense unit; **b** Constitution of dense connectivity with a 5-layer dense block; **c** A 3D-DenseNet with two blocks, the number of feature volumes was changed between blocks through transition layers

the layer. To boost the training against the vanishing gradients and improve the information flow within the network, the DenseNet [25] implements the connections from a layer to all its subsequent layers. We extended the idea of dense connectivity to 3D volumetric image processing tasks. In particular, x_l is defined as

$$x_l = H_l([x_0, x_1, \dots, x_{l-1}]), \tag{1}$$

where $x_0, x_1, \dots, x_{l-1}$ are 3D feature volumes produced in preceding layers, [...] refers to the concatenation operation. Figure 3a illustrates a dense unit. Composite function $H_l(\cdot)$ consists of three operations: a batch normalization (BN) to reduce internal covariate transform [26], a rectified linear unit (ReLU) to accelerate training process, and spatial convolution with k $3 \times 3 \times 3$ convolution kernels to generate 3D feature volumes. The basic framework of a 3D dense block is shown in Fig. 3b. A dense unit is regarded as one layer in a dense block and each layer is connected with all subsequent layers directly. With this dense connection mechanism, feature utilization become more effective and fewer feature increments are added to each layer than traditional CNNs. Therefore, the network are very narrow and has fewer parameters.

Bottleneck layers and transition layers are also used to reduce parameters. $1 \times 1 \times 1$ convolution is employed as bottleneck layer to reduce the input feature volumes, before convolution layer [27, 28]. After the mechanism of the bottleneck layer, multichannel feature volumes are fused and only a small set of feature volumes were added to the next layer while the preceding features are remained. Transition layers were also introduced to further improve model compactness with the hyperpa-

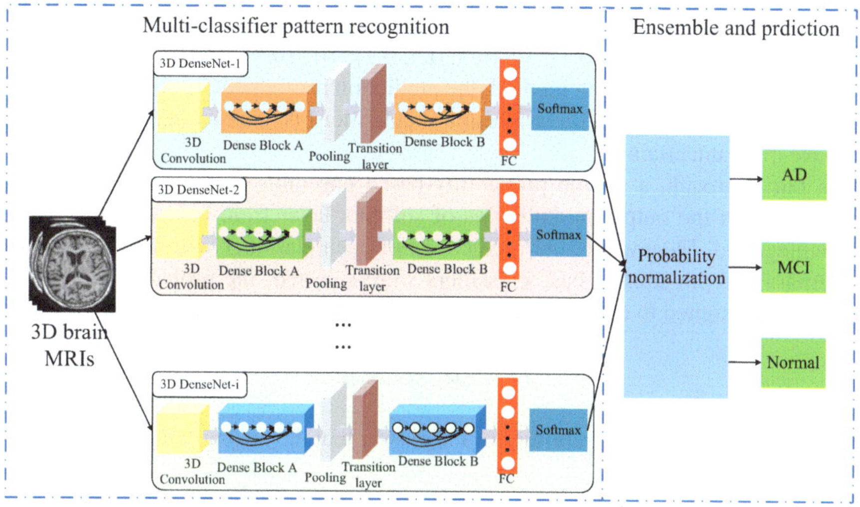

Fig. 4 Architectures of proposed ensemble 3D-DenseNet framework for AD and MCI diagnosis

rameter theta controlling the degree of compression. For a dense block that contains m feature volumes, the output feature volumes of the following transition layer decrease to $\lfloor \theta m \rfloor$, where $0 < \theta \leq 1$. Therefore, the 3D-DenseNet layers become very narrow and require fewer parameters than the traditional network, but it can perform well by making the best use of features through dense connections. To further eliminate redundancy and improve the feature expression performance of the model, we introduce dropout between the pooling layer and linear layer. The 3D-DenseNet with two dense blocks is illustrated schematically in Fig. 3c.

As mentioned above, the 3D-DenseNets with different hyperparameter sets appeared in various architectures. We demonstrated that the performance of 3D-DenseNet was sensitive to its hyperparameters through extensive experiments in Section IV. So training with different hyperparameters can adjust the instability of the base 3D-DenseNets and enhance their diversity. Based on extensive experimental results with varying hyperparameter sets, we generated base networks with different structures by changing hyperparameters randomly around the optimal value. All base 3D-DenseNets work independently and output the class probabilistic score by a softmax layer. We fused their outputs by the probability-based fusion method. The proposed ensemble model is shown in Fig. 4.

In the traditional majority voting method, the prediction results of most classifiers are used as the final prediction labels. Each classifier is independent and the error rates between different classifiers are irrelevant so that the performance of the ensemble model is better than a single classifier. But for multi-classification tasks, this method may not be very effective. Single classifiers perform well on most subjects, but for some subjects which are difficult to classify, the error rates will increase due to the uncertainty among multiple categories. For example, three classifiers are considered, the output probabilities of softmax lay for $\{AD, MCI, Normal\}$ are I: $\{0.8, 0.1, 0.1\}$, II: $\{0.4, 0.5, 0.1\}$, III: $\{0.3, 0.4, 0.3\}$, respectively. Based on the majority voting method, the prediction result is MCI. But it is not completely correct since the prediction result of classifier I is more credible while II and III have more uncertainty.

In our approach, a simple probability-based ensemble method was employed [29], in which the output probabilities of softmax layer from base classifiers will be reintegrated. Meanwhile, the predictions of each classifier will not be ignored. In ternary classification, i base classifiers were selected; the probabilities of 3D-DenseNet$_i$ assigned to categories on testing set were

$$P^i = (\alpha_1^i, \alpha_2^i, \alpha_3^i), \tag{2}$$

where α_j^i indicates the probabilities of the class j that the testing sample belongs to. We normalized the P^i by

$$P^i = \frac{P^i}{\max[\alpha_1^i, \alpha_2^i, \alpha_3^i]}, \tag{3}$$

where $max[\alpha_1^i, \alpha_2^i, \alpha_3^i]$ is the maximum element value of P^i. When outputs of m base 3D-DenseNets have been computed, the final class label was determined by the proposed fusion model as follows:

$$\mathrm{y} = \arg\max\left(\prod_{i=1}^{m}\alpha_1^i, \prod_{i=1}^{m}\alpha_2^i, \prod_{i=1}^{m}\alpha_3^i\right). \tag{4}$$

4 Experiments

4.1 Data and Implementation

In this section, we employed 833 MR subjects from ADNI to evaluate the proposed framework, including 221 AD subjects, 297 MCI subjects, and 315 Normal control subjects. And 10-fold cross-validation method was utilized to test the model performance. The original sample set was randomly partitioned into 10 equal-sized subsamples, a single subsample was retained as the validation data for testing the model, and the remaining 9 subsamples composed training set. The cross-validation process was then repeated 10 times, with each of the 10 subsamples used exactly once as the validation data. A ternary classifier (AD versus MCI versus Normal) and three respective binary classifiers (AD versus Normal, AD versus MCI, and MCI versus Normal) are used to report the classification results. The subjects selected from the same participant were forbidden from appearing in both the training set and the testing set. All the experiments were performed on a system with NVIDIA Tesla P100 GPU.

Table 1 Confusion matrix of binary classification

	Predicted Positive (Class A)	Predicted Negative (Class B)
Actual positive (Class A)	True positive (TP)	False negative (FN)
Actual negative (Class B)	False positive (FP)	True negative (TN)

Table 2 Confusion matrix of ternary classification

	Predicted Class A	Predicted Class B	Predicted Class C
Actual Class A	True A (T_A)	False AB (F_{AB})	False AC (F_{AC})
Actual Class B	False BA (F_{BA})	True B (T_B)	False BC (F_{BC})
Actual Class C	False CA (F_{CA})	False CB (F_{CB})	True C (T_C)

4.2 Experimental Steps and Evaluation

A 3D-DenseNet was selected as a base classifier for comparison with the ensemble method. The base 3D-DenseNet was trained and a series of experiments were conducted to choose optimal hyperparameters. Subsequently, some 3D-DenseNets were generated by varying primary hyperparameters around the selected optimal values randomly. Then comparison of the ensemble method and base classifiers was conducted to prove superiority of the ensemble method.

The performance of the classifiers can be interpreted from the confusion matrices, which record model performance across categories. The confusion matrices of binary and ternary classification problems are shown in Table 1 and Table 2, respectively. The average of 10-fold cross-validation was regarded as the final results.

The performance of each model is defined as follows:

(a) Accuracy that indicates the proportion of correctly classified subjects among the whole subset,

$$Accuracy_{bin} = \frac{TP + TN}{TP + TN + FP + FN}, \tag{5}$$

$$Accuracy_{ter} = \frac{T_A + T_B + T_C}{T + F}. \tag{6}$$

(b) Precision that quantifies the proportion of samples correctly classified among the classification,

$$Precision_{bin} = \frac{TP}{TP + FP}, \tag{7}$$

$$Precision_{ter-class A} = \frac{T_A}{T_A + F_{BA} + F_{CA}}. \tag{8}$$

(c) Recall is the fraction of relevant instances that have been retrieved over the total amount of relevant instances,

$$Recall_{bin} = \frac{TP}{TP + FN}, \tag{9}$$

$$Recall_{ter-classA} = \frac{T_A}{T_A + F_{AB} + F_{AC}}. \tag{10}$$

(d) F1-score considers both the precision and recall and evaluate the model performance synthetically,

$$F_1 - score = \frac{2 \times Precision \times Recall}{Precision + Recall}. \tag{11}$$

4.3 Parametric Analyses

A series of experiments were conducted to analyze the performance of the 3D-DenseNet with different hyperparameters sets including depth, growth rate, and compression factor. We report the testing accuracy and errors among the 10-fold cross-validation via boxplots which represent the interquartile ranges. The boxes indicate the quartiles of the dataset while the whiskers extend to show the points that are determined to be outliers using a method which is a function of the interquartile range.

Analysis of growth rate. The hyperparameter k was referred to as the growth rate of the network which indicates the number of new feature volumes increased at each layer. As shown in Fig. 5, the accuracy of classifiers vary significantly according to different k. The AD/MCI classifier obtains state-of-the-art result with $k = 15$ while MCI/Normal with $k = 12$. The model with $k = 9$ is sufficient to classify AD and Normal. As for the ternary classification task, state-of-the-art accuracy is achieved when $k = 24$. The model obtains poor accuracy with a small growth rate because essential features for classification are not fully extracted. Relatively large growth rate can boost the performance by introducing more feature volumes. However, overlarge growth rate may decrease the accuracy because the complicated model is deficiently trained with limited training data. What is more, the model with a larger growth rate has a smaller range of errors, the explanation for this is that complex structure with more parameters can improve the performance of the model.

Analysis of depth. The depth is referred to as the total number of layers of all blocks in the 3D-DenseNet. To investigate the impact of depth on the accuracy, the model with different depths was trained for experiments. The comparison of model performance with different depths is shown in Fig. 6. State-of-the-art accuracy can be obtained to categorize AD/Normal with depth = 15 while other classifiers obtain optimal accuracy with depth = 20. The mean accuracy of AD/MCI is lower than

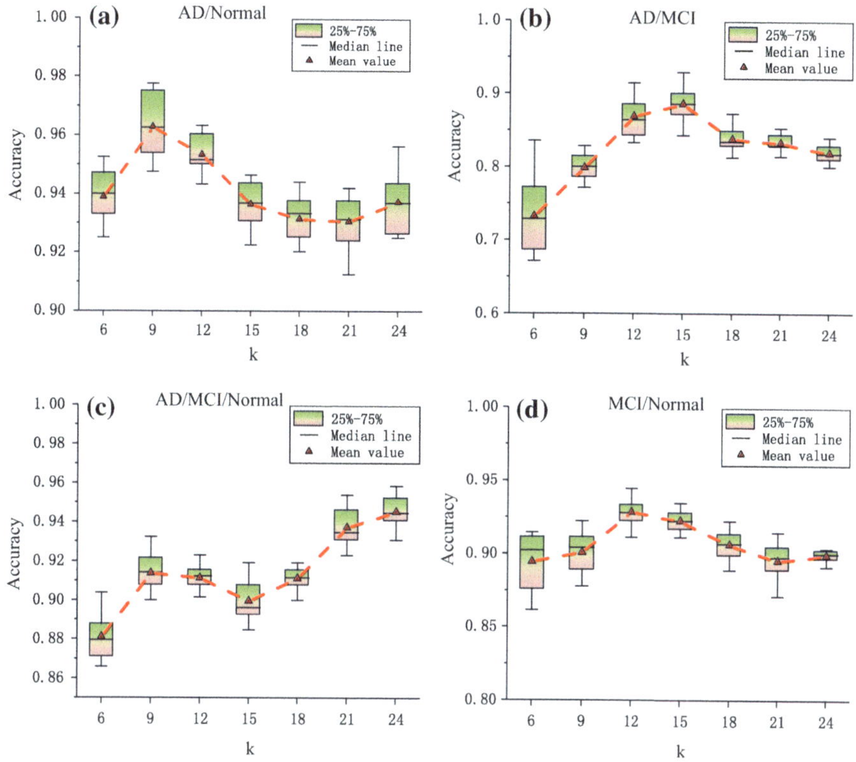

Fig. 5 Comparison of different growth rates

the others, because AD is usually developed from MCI and the anatomical shape variations captured from MRIs are not obvious enough to identify categories. Similar to the growth rate, the network with few layers cannot express features adequately, so increasing depth properly can improve the classification accuracy. But the network with overlarge depth may obtain poor accuracy, because parameters may not be fully trained due to limited dataset.

Analysis of compression factor. The θ is referred to as a compression factor, which indicates the degree of feature reduction in transition layers. Figure 7 shows that the variation of compression factor has a prominent impact on classification accuracy. The MCI/Normal and ternary classifiers obtain optimal accuracy with a medium compression factor. The accuracy of AD/MCI declines as θ increase. And the AD/Normal obtains optimal accuracy when the model is compressed with larger θ. Simplified network structures can ignore irrelevant features for diagnosis of dementia disease and reduce overfitting to some extent. But excessive compression of the network can lead to inadequate expression of features and thus reduce the accuracy of the model.

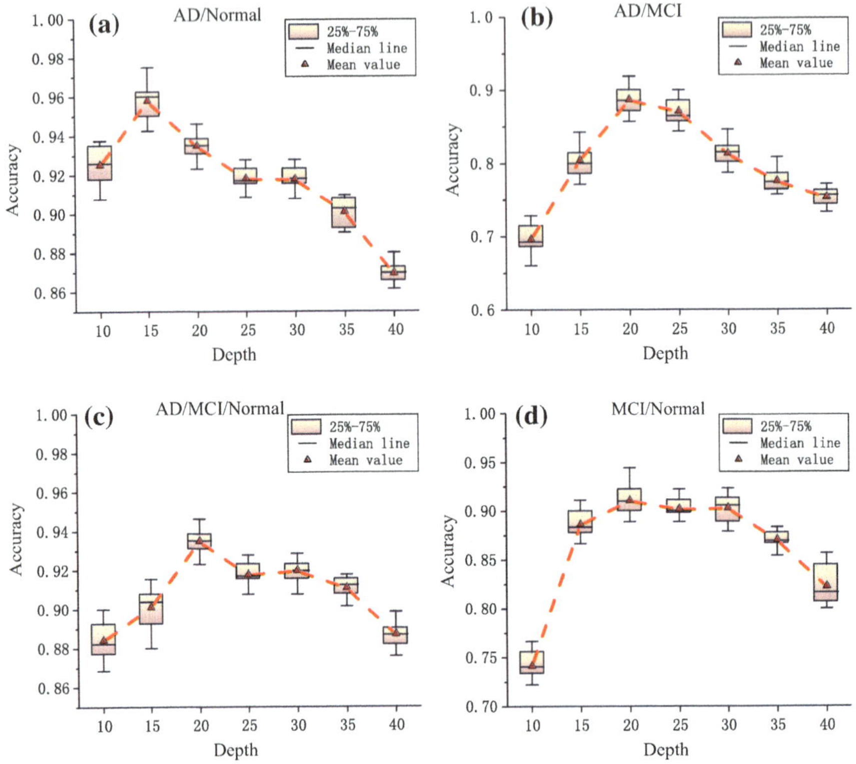

Fig. 6 Comparison of different depths

Analysis of the optimization methods. To further accelerate the model converging and improve its performance, BP arithmetic is mended by appending optimization methods. As we can see in Fig. 8, the model optimized by the momentum method achieved better accuracy than models with other optimizers in all classification tasks. One explanation for this is that current gradient is the accumulation of previous momentum. Larger values of momentum factor accelerate parameters update and help the model get rid of the local minimum when gradient decreases to zero.

Analysis of the model performance with different amount of training data. In order to analyze the effect of different parameters on the performance of the model, different number of samples were used for ternary classifier training. Figure 9 shows the mean accuracy with different amount of training data, where the models with different fixed hyperparameters were compared. Accuracy of the model with a fixed growth rate reduced rapidly with reduction of training samples while the models with fixed depth and compression factor change gently. @@So the depth is more insensitive to the amount of training data. In 3D-DenseNet, the growth rate focuses on the number of features which are sufficient to distinguish categories while the depth emphasizes expressing features with appropriate layers and mining the differences

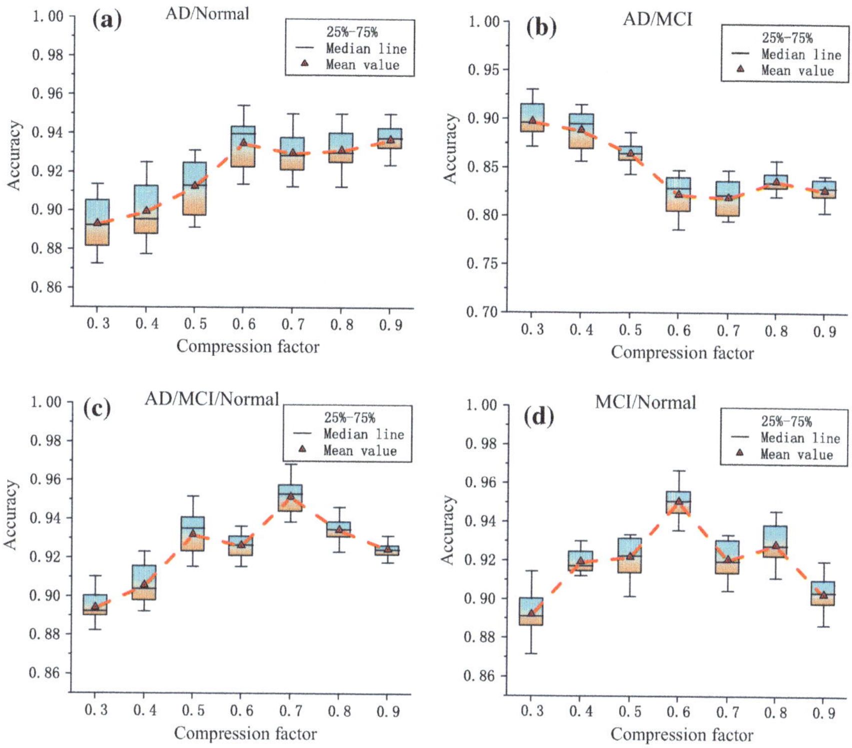

Fig. 7 Comparison of different compression factors

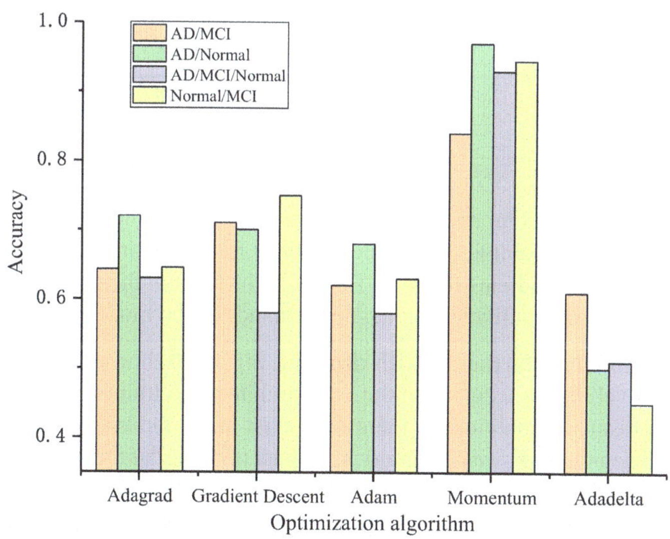

Fig. 8 Comparison of model performance with different optimization methods

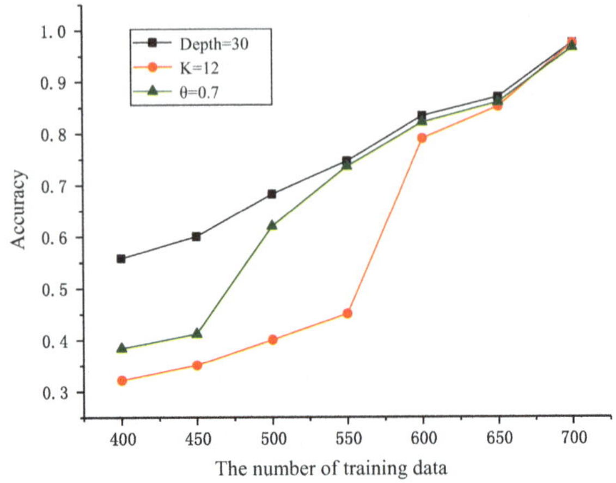

Fig. 9 Parameters' sensitivity to the number of training data

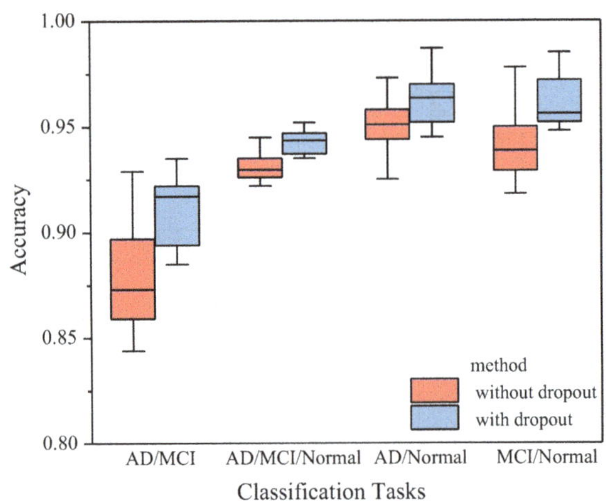

Fig. 10 Analysis of model performance with and without dropout

between features. And compression factors improve model compactness by reducing the amount of feature volumes. Dense connections reuse existing features and make the information and gradients transfer effectively throughout the network, so proper depth has an important influence on the performance of the model.

Analysis of model performance with dropout. Figure 10 shows the distribution of 10-fold cross-validation accuracy of the same architectures trained with and without dropout. As discussed previously, there may exist many redundant features in the output of the dense block, we decreased the impact of redundant features by dropout. Dropout reduced complex coadaptations of neurons and forced classifier to give up noise by dropping some units from the network with a certain probability temporarily. Therefore, the proposed model learned more pivotal features and prevents overfitting in this way.

Table 3 Comparison of parameter quantity and computation time with different network structure

Methods	Depth	Growth rate	θ	Parameters (M)	Time (h)
DenseNet-I	30	12	1	1.3	2.33
DenseNet-II	30	24	1	5.3	4.25
DenseNet-III	30	12	0.8	0.3	2.4
DenseNet-IV	30	12	0.5	0.2	2.5
DenseNet-V	50	12	1	4.4	6.4

Analysis of parameter quantity and computation time. In order to investigate the computational efficiency of the 3D-DenseNet, we compared parameter quantities and training times of different structures. These were computed on an NVIDIA Tesla P100 GPU and iterated for 150 epochs with a 0.01 initial learning rate. As shown in Table 3, the network with a larger growth rate have more parameters and consumes more time because there are more feature volumes increment at each layer. The narrow but deeper network also has higher time complexity due to it containing more layers in each dense block and extracting more abstract features. What is more, the compression of the transition layer can reduce parameters significantly and consume less memory, but cannot save computing time. This may because the channels of former collective knowledge were reintegrated, but features computing in each dense block cannot be simplified.

Analysis of parameter efficiency. The 3D-DenseNet with bottleneck structures is referred to as 3D-DenseNet-B while the 3D-DenseNet with dimension reduction is referred to as 3D-DenseNet-C and the 3D-DenseNet-BC denotes the model with both bottleneck and transition layers. To illustrate the parameter efficiency of variants of 3D-DenseNets, we train 3D-DenseNets with varying depths and growth rates on ADNI for ternary classification, and plot their test errors as a function of network parameters. The left plot in Fig. 11 shows that the 3D-DenseNet-BC, which can achieve a low error rate with relatively fewer parameters, is consistently the most parameter efficient variant of 3D-DenseNet. Multichannel feature volumes are fused by the 1 1 1 convolution layers, which are employed as bottleneck layers. And parameters are compressed by transition layers that lead to more compact models. The right plot in Fig. 11 shows that a 3D-DenseNet can achieve a much lower test error than 3D-CNN with the same number of parameters. Further, to achieve the same level of accuracy, 3D-DenseNet-BC only requires around a quarter of the parameters of 3D-CNN. Dense connection encourages feature reuse and less feature increment, which leads to more compact models.

4.4 Results

For each 3D-DenseNet, we initialized weights of the network randomly with a Gaussian distribution ($\mu = 0, \sigma = 0.01$). The initial learning rate was set to 0.01, and the poly learning rate policy was employed to update learning rate by multiply-

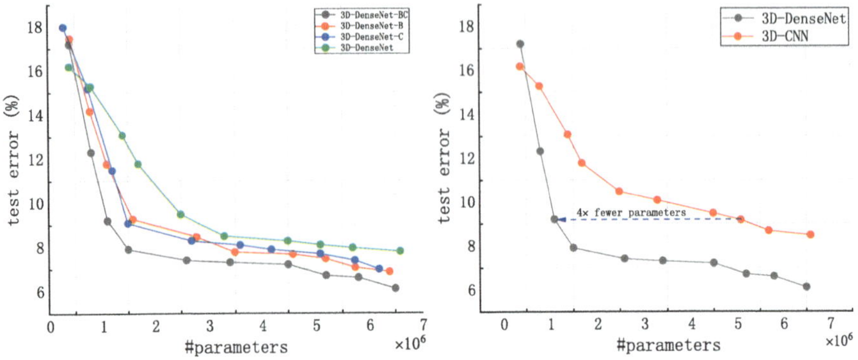

Fig. 11 Left: Comparison of the parameter efficiency on ternary classification. Right: Comparison of the parameter efficiency between 3D-DenseNet-BC and 3D-CNN. 3D-DenseNet-BC requires about a quarter of the parameters as 3D-CNN to achieve comparable accuracy

ing $(1 - \frac{iter}{max\_iter})^{power}$ along the training iteration. Momentum method (with batch size = 10, weight decay = 0.0005, and momentum = 0.9) was used to optimize training iteration. We carried out 1000 iterations for training and testing in each cross-validation. Finally, five different 3D-DenseNets were selected as base classifiers, and the variation ranges of accuracy among them are within 2%. The average of 10-fold cross-validation was regarded as the final results.

The experimental results are shown in Table 4. The best performance with accuracy of 97.52%, average precision of 97.13%, average recall of 97.0%, and F1-score of 97.1% is given by the proposed probability-based ensemble model, while the proposed 3D-DenseNet produce the accuracy of 94.77% and the majority voting method achieve the accuracy of 95.96%. From Table 4, the following findings can be given (1) The majority voting approach and the probability-based ensemble model can improve the model performance significantly. The fusion of multiple independent classifiers can reduce the error rate. (2) The proposed probability-based ensemble model outperformed the majority voting method. As mentioned above, this is mainly because the probability-based method can cumulate category probabilities of multi-base classifiers and make a prediction based on integrated information, rather than select the majority result directly.

In order to estimate the generalization capability of our proposed method, experiments were also conducted on three binary classification tasks (AD versus NC, AD versus MCI, and MCI versus Normal). Through the ensemble 3D-DenseNet method, encouraging accuracy results were obtained 98.83% for AD/Normal, 93.61% for AD/MCI, 98.42% for MCI/Normal, and 97.52% for AD/MCI/Normal. In order to display the performance of the model intuitively, the confusion matrices of the ensemble model on one of the random cross-validation are shown in Fig. 12.

Table 5 show the comparisons of the proposed model with previous methods on ternary tasks. Billones et al. [18] extracted 20 coronal slices from the sequence of MRI images and modeled each 2D slice separately. The modeling method based on 2D slices does not consider the correlation characteristics between slice sequences, so the classification performance is poor. The analysis of 3D MRI image can retain

Table 4 The performance of 3D-DenseNets and the ensemble model on testing set for AD/MCI/Normal

Model	Class	Accuracy	Precision	Recall	F1-score
Optimal 3D-DenseNet	AD	0.9477	0.9253	0.9696	0.9469
	MCI		0.9431	0.9325	0.9405
	Normal		0.9680	0.9578	0.9628
Average of the base classifiers	AD	0.9398	0.9104	0.9242	0.9172
	MCI		0.9425	0.9213	0.9317
	Normal		0.9578	0.9680	0.9628
Majority voting	AD	0.9596	0.9365	0.9402	0.9383
	MCI		0.9435	0.9526	0.9480
	Normal		0.9684	0.9703	0.9693
Probability-based ensemble model	AD	0.9752	0.9692	0.9545	0.9617
	MCI		0.9555	0.9662	0.9598
	Normal		0.9893	0.9893	0.9893

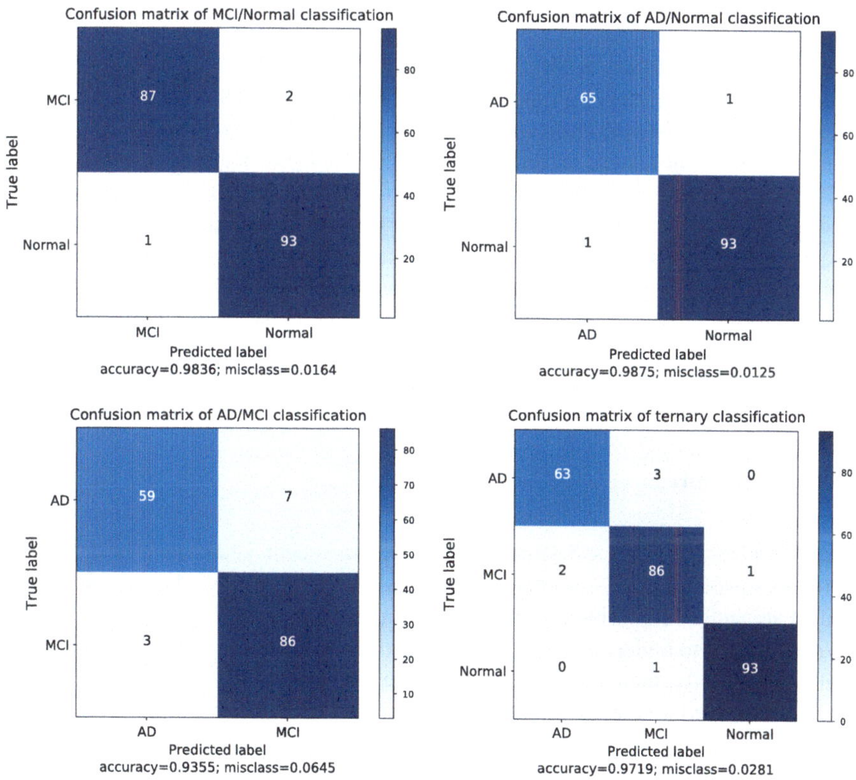

Fig. 12 Confusion matrix of four classifiers

Table 5 Accuracy of AD/Normal/MCI classification for different classification methods

Methods	Accuracy (%)
Billones et al. [18]	91.85
Hosseini et al. [19]	89.1
Payan et al. [21]	89.47
Cheng et al. [20]	87.15
3D-DenseNet	94.77
Majority voting	95.96
Proposed ensemble method	**97.52**

more spatial feature information, but the complexity of 3D image and 3D convolution operation increases. In order to ensure the computational efficiency of model training, Cheng et al. [21] divide 3D MRI image into several blocks, and extract feature information from each block by convolution operation, and then reorganize and combine modeling. This method can effectively extract local 3D feature information and simplify the operation, but ignores the correlation features between adjacent blocks. Hosseini et al. [19] and Payean et al. [20] performed convolution operations on the complete 3D MRI images. Key features for dementia diagnosis were automatically extracted by the 3D-CNN method, which improved the classification and diagnosis accuracy of the model. The 3D-DenseNet model proposed in this chapter introduces a dense connection mechanism on the basis of 3D-CNN to improve the gradient transfer efficiency and improve the feature utilization rate through feature reuse. Only a small feature increment is needed to achieve accurate pattern classification, which greatly reduces the amount of parameters and improves the operational efficiency. Therefore, the 3D-DenseNet model is superior to related methods in classification performance. Compared with the traditional majority voting method, the probabilistic ensemble learning method proposed in this chapter fully considers the sensitivity of each sub-classifier to the category and can reduce the prediction error more effectively. It integrates probabilistic ensemble learning and the 3D-DenseNet model to achieve better classification and diagnosis performance of dementia.

5 Conclusion

Deep learning has seen a dramatic resurgence in recent years, mostly driven by increases in computational power and the availability of massive datasets. Fields of health care and medicine have witnessed striking advances in the ability of computer-aided diagnosis to understand and manipulate data. But there are still many technical challenges. Firstly, the existing medical image datasets are relatively small in scale and lack of standardization in data acquisition and annotation process. Therefore, how to use limited datasets to construct an assistant diagnosis model with universal application value is a research hot spot of medical artificial intelligence in recent years. Secondly, medical images have the characteristics of high dimension and complex features. It is necessary to construct deeper network to extract effective features. However, the deepening of network layers has brought about a sharp increase in the number of parameters, and it is difficult to fully train with limited data. In addition, the deepening of the network layer makes gradient information gradually weaken and even disappear in the process of transmission, resulting in mode collapse.

In view of the above problems, we proposed a 3D densely connected convolutional network for early diagnosis of Alzheimer's disease. Compared with the traditional method based on 2D image slices, 3D convolution retains and extracts more key spatial feature information of brain MRI images, thus providing more feature basis for the classification model. Dense connection mechanism makes every layer of the convolutional network directly connected with each other and encourage feature reusing, which improves the transmission efficiency of feature and gradient information in the network. Through the extreme utilization of features, the feature increment of the convolution layer is reduced, so the network is relatively narrow, which greatly reduces the parameters of the network. In addition, through the mechanism of the bottleneck layer and transition layer, the network parameters are further reduced. The proposed model has the advantages of small parameters and high efficiency of feature transfer. The network can achieve deeper layers and avoid overfitting to a certain extent. We also developed a simple but effective probability-based ensemble learning method. The sub-classifier used to construct the integrated model is randomly adjusted near the optimal hyperparameters of 3D-DenseNet. By normalizing the predictive probability of each sub-classifier in softmax layer, the knowledge of each sub-classifier is integrated, instead of simply discarding some sub-classifiers through the majority selection principle like the traditional voting method. The advantage of this approach is that the sensitivity and specificity of each sub-classifier to the class samples are considered. Experiments verify the effectiveness of the probabilistic ensemble method, especially for multi-classification problems. The ensemble model achieved obvious boosting of accuracy than doing just the simple average of the network's predictions.

References

1. Alzheimer's Association et al., 2017 Alzheimer's disease facts and figures. Alzheimer's Dement. **13**(4), 325–373 (2017)
2. S. Li, O. Okonkwo, M. Albert, M.-C. Wang, Variation in variables that predict progression from MCI to AD dementia over duration of follow-up. Am. J. Alzheimer's Dis. **2**(1), 12–28 (2013)
3. R. Cuingnet, E. Gerardin, J. Tessieras, G. Auzias, S. Lehéricy, M.-O. Habert, M. Chupin, H. Benali, O. Colliot, A.D.N. Initiative et al., Automatic classification of patients with alzheimer's disease from structural MRI: a comparison of ten methods using the adni database. Neuroimage **56**(2), 766–781 (2011)
4. F. Falahati, E. Westman, A. Simmons, Multivariate data analysis and machine learning in alzheimer's disease with a focus on structural magnetic resonance imaging. J. Alzheimer's Dis. **41**(3), 685–708 (2014)

5. E. Moradi, A. Pepe, C. Gaser, H. Huttunen, J. Tohka, A.D.N. Initiative et al., Machine learning framework for early MRI-based alzheimer's conversion prediction in mci subjects. Neuroimage **104**, 398–412 (2015)
6. J.-Z. Cheng, D. Ni, Y.-H. Chou, J. Qin, C.-M. Tiu, Y.-C. Chang, C.-S. Huang, D. Shen, C.-M. Chen, Computer-aided diagnosis with deep learning architecture: applications to breast lesions in us images and pulmonary nodules in ct scans. Sci. Rep. **6**, 24454 (2016)
7. S.M. Plis, D.R. Hjelm, R. Salakhutdinov, E.A. Allen, H.J. Bockholt, J.D. Long, H.J. Johnson, J.S. Paulsen, J.A. Turner, V.D. Calhoun, Deep learning for neuroimaging: a validation study. Front. Neurosci. **8**, 229 (2014)
8. F.C. Ghesu, B. Georgescu, T. Mansi, D. Neumann, J. Hornegger, D. Comaniciu, An artificial agent for anatomical landmark detection in medical images, in *International Conference on Medical Image Computing and Computer-Assisted Intervention* (Springer, Berlin, 2016), pp. 229–237
9. W. Shen, M. Zhou, F. Yang, C. Yang, J. Tian, Multi-scale convolutional neural networks for lung nodule classification, in *International Conference on Information Processing in Medical Imaging* (Springer, Berlin, 2015), pp. 588–599
10. S. Wang, Y. Shen, C. Shi, P. Yin, Z. Wang, P.W.-H. Cheung, J.P.Y. Cheung, K.D.-K. Luk, Y. Hu, Skeletal maturity recognition using a fully automated system with convolutional neural networks, IEEE Access **6**, 29979–29993 (2018)
11. S.L. Risacher, A.J. Saykin, J.D. Wes, L. Shen, H.A. Firpi, B.C. McDonald, Baseline MRI predictors of conversion from MCI to probable AD in the ADNI cohort. Curr. Alzheimer Res. **6**(4), 347–361 (2009)
12. W. Cai, S. Liu, L. Wen, S. Eberl, M. J. Fulham, D. Feng, 3D neurological image retrieval with localized pathology-centric CMRGlc patterns, in *2010 17th IEEE International Conference on Image Processing (ICIP)* (IEEE, Piscataway, 2010), pp. 3201–3204
13. S. Liu, Y. Song, W. Cai, S. Pujol, R. Kikinis, X. Wang, D. Feng, Multifold Bayesian kernelization in Alzheimer's diagnosis, in *International Conference on Medical Image Computing and Computer-Assisted Intervention* (Springer, Berlin, 2013), pp. 303–310
14. D. Zhang, Y. Wang, L. Zhou, H. Yuan, D. Shen, A.D.N. Initiative et al., Multimodal classification of Alzheimer's disease and Mild Cognitive Impairment. Neuroimage **55**(3), 856–867 (2011)
15. F. Zhang, Y. Song, S. Liu, S. Pujol, R. Kikinis, M. Fulham, D. Feng, W. Cai, Semantic association for neuroimaging classification of PET images. J. Nucl. Med. **55**(supplement 1), 2029 (2014)
16. S. Liu, S. Liu, W. Cai, H. Che, S. Pujol, R. Kikinis, D. Feng, M.J. Fulham et al., Multimodal neuroimaging feature learning for multiclass diagnosis of Alzheimer's disease. IEEE Trans. Biomed. Eng. **62**(4), 1132–1140 (2015)
17. F. Li, L. Tran, K.-H. Thung, S. Ji, D. Shen, J. Li, A robust deep model for improved classification of AD/MCI patients. IEEE J. Biomed. Health Inform. **19**(5), 1610–1616 (2015)
18. C.D. Billones, O.J. L.D. Demetria, D.E.D. Hostallero, P.C. Naval, Demnet: a convolutional neural network for the detection of Alzheimer's Disease and Mild Cognitive Impairment, in *Proceedings of the 2016 IEEE Region 10 Conference (TENCON)* (IEEE, Piscataway, 2016), pp. 3724–3727
19. E. Hosseini-Asl, R. Keynton, A. El-Baz, Alzheimer's disease diagnostics by adaptation of 3D convolutional network, in *2016 IEEE International Conference on Image Processing (ICIP)*. (IEEE, Piscataway, 2016), pp. 126–130
20. A. Payan, G. Montana, Predicting Alzheimer's disease: a neuroimaging study with 3D convolutional neural networks (2015), arXiv:1502.02506
21. D. Cheng, M. Liu, J. Fu, Y. Wang, Classification of MR brain images by combination of multi-CNNs for ad diagnosis, in *Ninth International Conference on Digital Image Processing (ICDIP 2017)*, vol. 10420 (International Society for Optics and Photonics, Bellingham, 2017), p. 1042042
22. C.R. Jack, M.A. Bernstein, N.C. Fox, P. Thompson, G. Alexander, D. Harvey, B. Borowski, P.J. Britson, J.L. Whitwell, C. Ward et al., The Alzheimer's disease neuroimaging initiative (ADNI): MRI methods. J. Magn. Reson. Imaging **27**(4), 685–691 (2008)

23. M.W. Woolrich, S. Jbabdi, B. Patenaude, M. Chappell, S. Makni, T. Behrens, C. Beckmann, M. Jenkinson, S.M. Smith, Bayesian analysis of neuroimaging data in FSL. Neuroimage **45**(1), S173–S186 (2009)
24. M. Jenkinson, P. Bannister, M. Brady, S. Smith, Improved optimization for the robust and accurate linear registration and motion correction of brain images. Neuroimage **17**(2), 825–841 (2002)
25. G. Huang, Z. Liu, K.Q. Weinberger, L. van der Maaten, Densely connected convolutional networks, in *Proceedings of the IEEE Conference on Computer Vision and Pattern Recognition*, vol. 1, Issue 2, (2017), p. 3
26. S. Ioffe, C. Szegedy, Batch normalization: accelerating deep network training by reducing internal covariate shift, in *International Conference on Machine Learning* (2015), pp. 448–456
27. C. Szegedy, V. Vanhoucke, S. Ioffe, J. Shlens, Z. Wojna, Rethinking the inception architecture for computer vision, in *Proceedings of the IEEE Conference on Computer Vision and Pattern Recognition* (2016), pp. 2818–2826
28. K. He, X. Zhang, S. Ren, J. Sun, Deep residual learning for image recognition, in *Proceedings of the IEEE conference on computer vision and pattern recognition* (2016), pp. 770–778
29. G. Wen, Z. Hou, H. Li, D. Li, L. Jiang, E. Xun, Ensemble of deep neural networks with probability-based fusion for facial expression recognition. Cogn. Comput. **9**(5), 597–610 (2017)

Detecting Work Zones in SHRP2 NDS Videos Using Deep Learning Based Computer Vision

Franklin Abodo, Robert Rittmuller, Brian Sumner and Andrew Berthaume

Abstract Naturalistic driving studies seek to perform the observations of human driver behavior in the variety of environmental conditions necessary to analyze, understand, and predict that behavior using statistical and physical models. The second Strategic Highway Research Program (SHRP2) funds a number of transportation safety-related projects including its primary effort, the Naturalistic Driving Study (NDS), and an effort supplementary to the NDS, the Roadway Information Database (RID). This work seeks to expand the range of answerable research questions that researchers might pose to the NDS and RID databases. Specifically, we present the SHRP2 NDS Video Analytics (SNVA) software application, which extracts information from NDS-instrumented vehicles' forward-facing camera footage and efficiently integrates that information into the RID, tying the video content to geolocations and other trip attributes. Of particular interest to researchers and other stakeholders in the integration of the work zone, traffic signal state, and weather information. The version of SNVA introduced here focuses on work zone detection, the highest priority. The ability to automate the discovery and cataloging of this information, and to do so quickly, is especially important given the two petabytes (2PB) size of the NDS video data set.

F. Abodo (✉) · R. Rittmuller · B. Sumner · A. Berthaume
Volpe National Transportation Systems Center, 55 Broadway, Cambridge, MA 02142, USA
e-mail: franklin.abodo@dot.gov

R. Rittmuller
e-mail: robert.rittmuller@dot.gov

B. Sumner
e-mail: brian.sumner@dot.gov

A. Berthaume
e-mail: andrew.berthaume@dot.gov

M. A. Wani et al. (eds.), *Deep Learning Applications*, Advances in Intelligent Systems and Computing 1098,
https://doi.org/10.1007/978-981-15-1816-4_5

1 Introduction

The Federal Highway Administration (FHWA) created and continues to fund the Second Strategic Highway Research Program or SHRP2. The primary safety-related product of this program was the conduction of its large-scale naturalistic driving study (NDS), which collected driver- and vehicle-specific data from human drivers in a wide variety of real-world roadway and environmental conditions. Volunteer driver vehicles were outfitted with two external and two internal cameras, forward-facing radar, and a Data Acquisition System (DAS) that collected telemetry data from the vehicle's CAN bus(including steering angles, acceleration, etc.). Most data was collected between 2012 and 2015 by more than 3,500 drivers across six states, spanning over five million trips and resulting in more than one million hours of recorded video. A complementary project to the NDS was the creation of the Roadway Information Database (RID), which contains stationary characteristics of the portions of the U.S. roadway network over which the NDS drivers drove, such as the number of lanes, the type of turn lane at an intersection, the presence or absence of rumble strips on a highway segment, and so on. Since the inception of SHRP2, the presence or absence of work zones in a given NDS trip has been a desired piece of information. Prior efforts to conflate work zone occurrences with roadway information in the RID were not successful because they depended on 511 traffic data provided by participating state departments of transportation. 511 data (so-called because that number can be dialed to access traffic information via telephone) is only sparsely informative, indicating what segments of what highways had construction planned within a given time period. Whether actual construction equipment was present on the particular segment of highway over which a volunteer driver drove and at the time he or she drove is not an answerable question. In some cases, researchers have used 511 data to identify trips and their accompanying videos that supposedly contained a work zone, only to find themselves manually skimming through videos, sometimes finding what they were looking for and other times not [1]. Further, one of the six participating states did not manage to supply 511-level data about work zones, making extraction of events from the video the only option for those trips. Here, we present the SHRP2 NDS Video Analytics (SNVA) software application, which aims to perform a complete and accurate accounting of work zone occurrences in the NDS video data set. To achieve this, SNVA combines a video decoder based on FFmpeg, an image scene classifier based on TensorFlow, and algorithms for reading timestamps off of video frames, for identifying the start and end timestamps of detected work zone events, and for exporting those events as records in CSV files. We organize the presentation of SNVA as follows: Sect. 2 presents related work; Sect. 3 discusses the motivations and methods behind our choice of deep learning framework and models; Sect. 4 details our approach to the joint development of the work zone detection model and the data set used to train it; and Sect. 5 describes our choice of hardware and software components and highlights efforts made to optimize the video processing pipeline. In Sect. 6 we present the expected future directions of SNVA development, and we conclude in Sect. 7.

2 Related Work

A number of efforts to extract knowledge from SHRP2 video using machine learning and computer vision techniques have been made prior to this work, with some succeeding as proofs-of-concept not originally intended for data set-wide applicability. [2] surveys some such projects funded by FHWA that performed vehicle and pedestrian detection, scene segmentation, traffic signal state detection, head, torso, and hand pose detection, and facial feature detection. In [3], the techniques used in several projects are discussed, including Haar Cascades, [4], for detecting vehicles and vehicle lights, Histogram of Oriented Gradients, [5], for vehicle detection, and convolutional neural networks, [6], for detecting the presence of a front-seat passenger. The authors of [7] used handcrafted features and off-the-shelf software to detect and track faces using the driver-facing camera.

Some reasons why earlier projects did not advance beyond the proof-of-concept phase include (1) limited training data, (2) the use of computer vision techniques that predate the second neural network renaissance, (3) the use of neural networks pre-trained on publicly available data sets with no additional transfer learning performed, and (4) slow processing speeds. SNVA addresses each of these issues primarily by taking advantage of deep learning architectures and software frameworks made available more recently than those used in earlier efforts.

3 Deep Learning Framework and Architecture Selection

3.1 Deep Learning Framework Selection

At the instantiation of the SNVA project, TensorFlow (TF) [8] was identified as the deep learning framework most likely to contribute the most to the project's success. We based this decision on two main factors. First, the apparent level of development and maintenance support as indicated by (1) the framework's popularity within machine learning research and practitioner communities as an open source tool, and (2) the framework's use in large-scale software applications by its creator Google. And second, the high-level API TensorFlow-Slim (TF-Slim), which was observed to include (1) useful demonstration code that could accelerate the team's learning and use of the framework, and (2) implementations of many CNN architectures accompanied by weights and biases pre-trained on ImageNet [9] for use in transfer learning [10].

3.2 CNN Architecture Selection

Convolutional neural networks (CNNs) have a widely demonstrated ability to apply to one task the weights and biases that were optimized for application to a different task. This is particularly true for tasks involving natural images such as those found in SHRP2 NDS video data. Confident that an off-the-shelf CNN would prove suitable for the scene detection task, we set out to compare seven architectures for their in-sample test performance, inference speed and GPU utilization: InceptionV3 [11], InceptionResnetV2 [12], MobilenetV1 [13], MobilenetV2 [14], NASNet-Mobile [15], ResnetV2-50 and ResnetV2-101 [16]. The similarly well known VGGNet [17] and AlexNet [6] architectures were not considered because of their low ratios of accuracy to number of parameters [15].

3.2.1 Validation Metric Selection

In order to compare models learned using different architectures against one another during testing, and also against themselves during training for the purpose of early stopping, a validation metric is required. One simple and popular metric is accuracy: the ratio of correctly classified samples to the total number of samples. If we assume that class representation in the training and target data sets is imbalanced (e.g. the ratio of work zone scenes to non-work zone scenes is very low), then accuracy becomes an unreasonable metric. In a pathological example, if the ratio of work zone to non-work zone samples were 1:19, then a model could assign 100% of test set samples to the not-work zone class and be 95% accurate in spite of not detecting a single work zone. With this in mind, we extended the TF-Slim demo code, which originally only measured accuracy, to add the following performance measures: precision, recall, F_1, $F_{0.5}$, F_2, true and false positives and negatives, and total misclassifications. While all of the aforementioned metrics were used to develop an intuition about each model's performance, $F_{0.5}$ was ultimately chosen as our single target measure. Recall that F-measures integrate precision and recall into a single metric, which is convenient when both measures are valuable. In the F_β formulation of the F-measure, [18], setting $\beta < 1$, $\beta > 1$, or $\beta = 1$ assigns more weight to precision, more weight to recall, or equal weight to precision and recall, respectively. In our use case, it is more important that the scene detector be correct when it claims to have discovered a work zone than to discover all existing work zones. We assume that the NDS data set is sufficiently large that some detections can be sacrificed for the benefit of relieving researchers of time wasted skimming through video clips that do not contain work zones. And so, we set $\beta = 0.5$.

3.2.2 Qualitative Filtering of Troublesome Samples

The quality of SHRP2 video data can vary on a number of dimensions. Camera or other hardware malfunctions can lead to discontinuities in video frames, completely noisy frames, completely black frames, frames in which the scene is rotated or frames that are out of focus, as illustrated in Fig. 1. Work zone scene features of interest may be too distant to label confidently. Rain and other environmental impacts on vehicle windshields may distort features of interest, making them look identical to

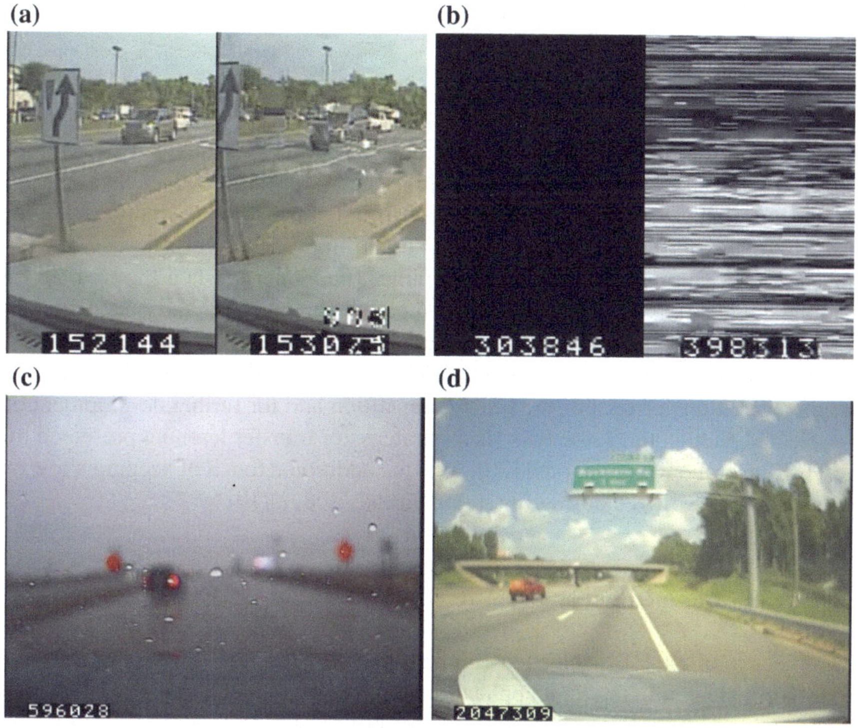

Fig. 1 **a** The left frame is eligible for inclusion in the training set, but its successor is distorted and would thus be excluded from consideration. **b** Both the left and right frames' entire source videos would be excluded from training and validation sets. **c** Although the camera's focus is on the raindrops on the windshield, leaving the background blurry, this frame would be included in the training set with a label of *warning sign* because the signs are sufficiently close to the vehicle not to be mistaken for background objects. The same signs at a slightly greater distance would easily not qualify. **d** This frame is also blurry because the camera is focused on a foreground object, but it would be included in the training set with a label of *not work zone* because it is obvious that no work zone equipment is present in the scene

features normally observed in background scenes. In some cases, we were required to exclude entire videos from consideration, but in most cases, a few frames here and there were set aside. Only frames that could be labeled without hesitation were included in the data set, with the hope that the absence of excluded frames during training would lead the model to classify frames similar to them with low confidence at test time. If the assumption were to hold, a smoothing algorithm could be applied to perturb low confidence frames to match the class of their highly confident neighbors, potentially correcting misclassifications caused by the very exclusion of such frames during training. This exact behavior was observed in a handful of randomly sampled outputs during the testing of the SNVA application.

3.2.3 Transfer Learning Using Weights and Biases Pre-trained on ImageNet

Conscious of the seemingly infinitesimally small amount of training data available to us, we only considered CNN architectures for which weights and biases pre-trained on the ImageNet 2012 Large-Scale Visual Recognition Challenge data set [9] were available for download. The transferring of such weights and biases from one task to another has been demonstrated to aid in prediction tasks across a wide range of applications and scientific disciplines, particularly when the number of training samples is very low [19]. We compared training time and out-of-sample test performance for several CNNs initialized using random weights versus pre-trained weights and found that all showed better performance in both cases when weights were transferred. For the CNN competition and for further development of the selected architectures, the two-phase strategy for transfer learning presented by the TF-Slim authors was adopted [20], with the additional touch of keeping a few of the earliest layers frozen during fine-tuning as advised in [10].

3.2.4 CNN Competition and Results

The objective of the CNN selection competition was to identify the single best candidate for inclusion in the final application. Architectures were compared using in-sample $F_{0.5}$ scores, inference speed in frames per second, and GPU core and memory utilization. In our experiments, we identified MobilenetV2 as the most suitable candidate because of its combination of highest inference speed, lowest memory consumption, and relatively high $F_{0.5}$ measure. The low memory consumption is of particular value because it permits either a large batch size or the concurrent assignment of multiple video processors to a single GPU. The competition results for all CNNs is presented in Table 1.

Table 1 CNN architecture performance comparison

Architecture	$F_{0.5}$	FPS	GPU (%)	GPU memory (MB)	Batch size	Steps (K)
InceptionV3	0.971	783	96	8031	32	47.4
InceptionResnetV2	0.957	323	96	7547	64	41.9
MobilenetV1	0.960	1607	**91**	8557	32	45.7
MobilenetV2	0.968	**1615**	94	**2413**	32	45.5
NASNet-Mobile	0.964	1211	98	2459	128	45.8
ResnetV2-50	**0.972**	1000	98	8543	64	46.7
ResnetV2-101	0.931	645	98	8543	128	46.4

4 Data Set Construction and Model Development

In this section, we describe the methods used to jointly develop the selected model defined in the previous section together with the data set used to train, validate, and test that model. Specifically, we discuss the processes by which we (1) determined what sources of data should contribute to data set construction, (2) defined a policy for excluding "unreasonable" data samples, (3) selected the CNN architectures that would compete for use in the final version of SNVA and the deep learning framework in which those architectures would be implemented, and (4) jointly developed the selected and training set.

4.1 Data Source Selection

Because the SHRP2 NDS video data was collected using homogeneously instrumented vehicles, we expected the SNVA application to target videos that were consistent in quality and characteristics (e.g. resolution, camera focal length and other intrinsic properties, et cetera). In turn, we limited our sources for data set construction to the NDS videos themselves, assuming that images from publicly available data sets that happened to contain construction features would be too out-of-sample to be useful in classifying the target distribution of scenes.

A total of 1344 videos containing 31,535,862 frames were made available for use as sources for data set construction, including training, validation, and test subsets. By manual inspection, the videos were observed to contain a variety of environmental scenes and features such as light and heavy rain, snow and fog, sunlight in front of, above and behind the subject vehicle, dusk, dawn, and nighttime scenes, and highway and city scenes. This variety gave us confidence that our data source scene distribution was representative of the target scene distribution, in spite of constituting less than 0.0001% of the estimated total number of frames in the target data set. Examples of the variety of scenes are presented in Fig. 2.

4.2 Active Learning via Uncertainty Sampling

Active learning is a set of semi-supervised techniques used to reduce the cost of machine learning model development. While there exist many varieties of active learning, all share the common objective of minimizing the number of training samples that require hand-labeling by a human, while still producing a model that meets inference performance requirements. For our purposes, we adopt a simple and commonly used method based on uncertainty sampling [21]. In this approach, a model is initially trained on a small hand-labeled training set of "seed" samples, then used predict the classes of the remaining unlabeled samples. Samples for which the model

Fig. 2 A variety of environmental conditions were present in the small subset of videos used for model development. **a** Clear-skied daytime with the sun behind the camera. **b** Nighttime with the subject vehicle's headlights illuminating construction drums. **c** Rainy daytime with the camera correctly focused on distant objects. **d** Dusk with the sun in front of the camera

is most uncertain in its label prediction (e.g. for which the probability distribution over classes is closest to uniform) are assumed to be the most informative to the model and are selected for inclusion in the next round of training, followed again by inference. This procedure is repeated until either (1) financial or human capital is exhausted, or (2) the model's performance converges (e.g. its lowest confidence prediction is above some desired threshold, or the number of uncertain samples per round stops decreasing). One can think of these most uncertain samples as being most informative to the model because their inclusion in the next round of training would adjust the model's decision boundaries the most. Seen from another perspective, if a model is confident in its prediction of an unlabeled sample's class, then adding that example to the training set will not contribute to improving the model's performance (assuming the prediction is correct) because it would not likely adjust the model's decision boundaries.

Of course, when the model makes highly confident predictions about the wrong class, then it would be beneficial to include the affected samples in the next round of training. This point raises a dilemma; the only way to observe these misclassifications is to inspect all predictions and not just the uncertain ones, which negates the premise of active learning entirely. The resolution to this dilemma is another assumption; that if we focus on hand-labeling only the uncertain samples, eventually the model will learn enough from them that it will either (1) correct its highly confident misclassifications or (2) decrease its confidence in the wrong class enough so that a human ends up labeling the sample directly as part of the routine active learning process. This second behavior was observed during our labeling effort, but not rigorously studied.

There remains a question of what probability threshold should mark the boundary between certain and uncertain class predictions. To begin to develop an answer, consider that in the most extreme case, no matter how many classes exist in a given classification problem, no more than one class can have confidence greater than 0.5. Thus, one of the uncertainty boundaries should be 0.5. Anything below this is definitely in need of a human-assigned label. For the upper bound, above which the model is considered certain in its prediction, the closer that threshold is to 0.5 or 1.0, the lower or higher the number of examples proposed for hand-labeling will be, respectively. In the absence of any analytic method for determining the value of the upper threshold, we take a stratified approach and define five ranges (above 0.5) within which data points may be binned: (0.5, 0.6], (0.6, 0.7], (0.7, 0.8], (0.8, 0.9], and (0.9, 1.0]. Following this approach, the amount of effort devoted to hand-labeling can be determined dynamically as a function of the number of samples in each bin. At a minimum, every sample in the [0.0, 0.5] is automatically labeled. The decision to hand-label points in the next higher bin can be made one step at a time. In our application of this strategy, we started with a seed set of 100,000 frames selected from eight videos. After the first round of training and inference over unlabeled samples, we selected an additional 350 from our total of 1314 videos to serve as sources for data set construction. Among these 350 videos, 50,000 frames were binned in the range [0.0, 0.5] and we selected those for labeling. For the second round, we were able to expand the range to include (0.5, 0.6] as only 30,000 were contained therein. We are at the time of this writing continuing to increase the training set size and improve the model while beta versions of the SNVA application are being tested in the deployment environment at VTTI. Active learning is particularly attractive for our use case and critical to our success for two reasons. First, because work zone features are expected to occur in the target data set relatively infrequently, a simple uniform random sampling of video frames for inclusion in the training set would likely (1) ignore useful work zone-containing frames, and (2) result in time wasted labeling frames with redundant information content. Second, because the number of available unlabeled frames approaches 2^{25}, discovering all work zone-containing frames by exhaustively skimming through videos is not feasible.

5 SNVA Application Design and Development

5.1 *Core Software Components*

5.1.1 TensorFlow and the TF-Slim Image Classification Model Library

The TensorFlow framework, together with the communities internal and external to Google that support it, is the primary enabler of this work. The public availability of quality demonstration code by way of the TF-Slim image classification library [20] accelerated experimentation with and development of neural network models substantially. In addition to pre-trained models, the library included Python scripts for the training and evaluation of models, the creation of TFRecord-based datasets, and the conversion of large model checkpoint files into compact, constant operation-only protobuf files that are optimized for inference. We were able to easily extend the code to support project-specific needs such as

1. the incremental training data set construction method used in active learning,
2. the augmentation of existing architecture implementations to support the NCHW format,
3. the addition of command-line parameters to make the running of multiple training and evaluation scripts concurrently across multiple GPUs and CPUs convenient, and
4. the addition of channel-wise standardization based on data set-level statistics as an optional preprocessing function.

Another surprisingly useful component of the TF ecosystem was Tensorboard, a visualization tool that helped us monitor training and evaluation, compare the performance of various architectures during the competition is mentioned in Sect. 3, and study and understand the structure and running state of TF-Slim models.

5.1.2 The FFmpeg Video/Audio Conversion Program

The decoding of the MPEG-4 formatted videos into raw bytes for ingestion by analyzers was performed using FFmpeg. The program was also used to extract individual frames from videos and save them to disk for use in model development.

5.1.3 The Numpy Scientific Computation Library

The Numpy scientific computation library was exercised heavily in the algorithms that process video frame timestamps and that apply weighted averaging to smooth class probability distributions output by SNVA's models. The library's support for broadcasting and vectorization sped up operations noticeably when used in place of the naive/intuitive implementations that preceded them.

5.1.4 The Python Programming Lanugage

The SNVA application was implemented entirely in Python, making it easy to integrate the software components used. TF's model development API is written in Python, and the FFmpeg binary is easily invoked and interacted with using Python's subprocess module. Given Python's popularity in the data science community, we expect its use to help make this project accessible to would-be beneficiaries inside and outside of the authors' affiliated organizations.

5.2 Video Frame Timestamp Extraction

The SHRP2 videos are one type of what is referred to as supplemental data. Supplemental data are stored on file systems and not directly integrated into the RID. To integrate information derived from these raw data into the RID, a process named *conflation* with a geospatial database named the Linear Referencing System (LRS) has been defined. To conflate work zone scene detections, allowing them to be localized using existing GPS information for a given trip, the beginning and end timestamps of detected scenes needed to be identified. Because the RID did not already contain a mapping from frame numbers to timestamps for every video, the only way to temporally localize work zone scenes in videos was to directly extract the digital timestamps graphically overlaid on video frames.

A pseudocode for the algorithm that performs this extraction is outlined in Algorithm 1: Lines 1 and 2 define the number of timestamps, l, and the maximum number of digits per timestamp, n, in the input array of timestamps, T. n is given by the ratio of the maximum timestamp width, w, and timestamp height, h, because each timestamp digit image is 16×16 pixels square. Lines 3 through 6 define M to be an array of binary image masks representing the ten Arabic numerals, and then prepare M to be compared for equality against each digit of each timestamp. The three color channels in T's third dimension are collapsed into one grayscale channel and then converted to binary black and white to match the format of M in line 7. In lines 8 and 9, T is reshaped to match the dimensions of M. Lines 10 and 11 test each digit in each timestamp for equality with all ten Arabic numeral masks and produce a 3D array of shape $l \times 10 \times n$, where one truth value exists for each numeral. Ideally, exactly one of the values will be True and the other nine False. Line 12 extracts matches using three 1D arrays to represent them, F, D, and P. Each array contains indices into one of the dimensions of the array output by line 11. Conveniently, the three arrays have a semantic interpretation, the values in F represent the frame number from which each timestamp was extracted, D contains the numerical values of timestamp digits, and the values in P represent the position at which those digits occurred in their timestamp. From these three arrays of integers, we can construct string representations of what were previously imaged.

Algorithm 1 ConvertTimestampImagesToStrings(T, h, w)

```
l ← LEN(T)
n ← w ÷ h
M ← GETTIMESTAMPDIGITMASKARRAY()
M ← TILE(M, n)
M ← TRANSPOSE(M, (0, 2, 1))
M ← RESHAPE(M, (l, n, h, h))
T ← BINARIZETIMESTAMPIMAGES(T)
T ← RESHAPE(T, (l, n, h, h))
T ← EXPANDDIMS(T, 1)
E ← EQUAL(T, M)
A ← ALL(E, (3, 4))
F, D, P ← NONZERO(A)
F_u, F_u^c ← UNIQUEWITHCOUNTS(F)
C_u, C_u^i, C_u^c ← UNIQUEWITHINDICESANDCOUNTS(F_u^c)
s ← SUM(C_u^c)
if s ≠ l then
    raise TIMESTAMPDETECTIONCOUNTERROR()
for i = 1 to LEN(C_u^i) − 1 do
    if C_u^i[i] < C_u^i[i − 1] then
        raise NONDECREASINGTIMESTAMPLENERROR()
D ← ASTYPE(D, UNICODETYPE)
S ← NDARRAY(l, INTEGERTYPE)
i_r ← 0
for i = 0 to LEN(C_u) − 1 do
    c_u ← C_u[i]                          ▷ length of each timestamp in batch j
    c_u^c ← C_u^c[i]                      ▷ number of c_u-length timestamps
    n_l^r ← c_u × c_u^c                   ▷ total digits spanning c_u^c timestamps
    i_l ← i_r                             ▷ left index into c_u-length timestamps in D
    i_r ← i_l + n_l^r                     ▷ right index into timestamps in D
    P_l^r ← P[i_l : i_r]                  ▷ timestamp-grouped digit positions
    P_l^r ← RESHAPE(P_l^r, (c_u^c, c_u))  ▷ timestamp-wise P_l^r
    P_l^r ← ARGSORT(P_l^r)                ▷ order positions increasingly
    O ← ARANGE(0, n_l^r, c_u)             ▷ define index offsets...
    O ← EXPANDDIMS(O, 1)                  ▷ ...into D for batch j
    P_l^r ← ADD(O, P_l^r)                 ▷ shift indices P_l^r by offsets O
    D_l^r ← D_l^r[i_l : i_r][P_l^r]       ▷ ordered c_u-length timestamps
    c_u^i ← C_u^i[i]                      ▷ first c_u-length timestamp index into D_l^r
    for j = c_u^i to c_u^i + c_u^c do     ▷ concatenate c_u digits...
        S[j] ← JOIN(D_l^r[j − c_u^i])     ▷ ...into one string
return S
```

Algorithm 2 BinarizeTimestampImages(T)

```
T ← AVERAGE(T, 2)          ▷ convert image to grayscale
t ← 128                    ▷ define binarization threshold
w ← [255]                  ▷ define white pixel value
b ← [0]                    ▷ define black pixel value
T ← WHERE(T >= t, w, b)    ▷ binarize image
return T
```

At this point, our ability to exploit Numpy's broadcasting feature becomes constrained due to potential variability in the number of digits that represent a timestamp within a single video (consider for example the transition from $t = 999965$ to $t = 1000031$ for $t \in T$). Continuing in the interest of minimizing computation, lines 13 and 14 prepare the algorithm to iterate over batches of timestamps sharing a common number of digits rather than over each timestamp one at a time. This is achieved by ordering the values in F to be monotonically non-decreasing, resulting in C_u, identifying the index of the first occurrence of each numeral in the arrangement, C_u^i, and separately the number of occurrences of each numeral in C_u^c.

Before proceeding, two quality control checks are performed between lines 15 and 20 to determine if any one of the timestamps could not be read due to some form of distortion in the corresponding video frame. The first looks for evidence of any missing timestamps. The second checks for evidence of an existing timestamp with one or more of its digits missing. If either check is satisfied, a 5% slower variant of the conversion algorithm that operates on each timestamp individually is run to identify the culprits and quality control them by synthesizing an artificial replacement timestamp. Because the resolution of link-level information in the LRS is much lower than the frame rate of videos, the error introduced by using synthetic timestamps is inconsequential. We do not present this alternative algorithm here.

With the extracted timestamps validated, conversion to the string representation can proceed. Line 21 converts the individual digits from integers into a string, and line 22 initializes a new n-dimensional array of length l in which the output strings will be placed. The variable i_r, which is used inside the loop starting on line 24 together with i_l to extract batches of the digits of equal-length timestamps, is initialized outside of the loop on line 23. For a detailed treatment of the string construction loop, please see the figure for Algorithm 1 starting at line 25.

Unfortunately, the operation that concatenates individual digit strings of length one into a single timestamp string of length c_u does not have a broadcast implementation in Numpy. The library simply calls the Python equivalent function. Thus, this $O(l)$ operation could not be optimized.

5.3 *Input Pipeline Optimization*

Particular attention was given to the design of SNVA's video processing pipeline, the core of the application. Much of the design was inspired by key features of the TF data ingestion API and by guidance in the TF documentation. Numpy and TF preprocessing are performed on the CPU in order to dedicate GPUs to the task of inference and maximize concurrency in the pipeline. An abstract graphical representation of the pipeline is presented in Fig. 3.

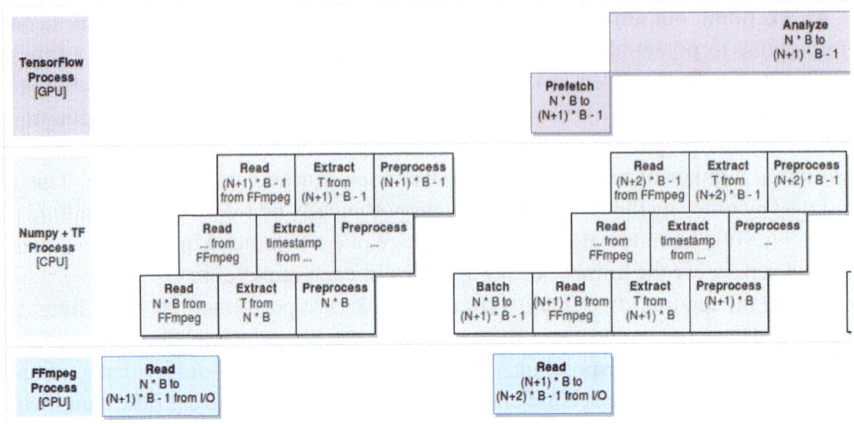

Fig. 3 Video Processing Pipeline: An abstract diagram describing the concurrent nature of the pipeline through which video data is ingested and processed. In the SNVA development environment, we achieved total GPU saturation while processing batches of frames within a given video, and only a 2–3 s delay between the last batch of one video and the first batch of the next

5.3.1 Generator Functions for Continuous Video Frame Streaming

The option of feeding the TF data ingestion pipeline using a generator function allowed us to pipe FFmpeg's output directly into TF, effectively handing the responsibility of optimizing the reading of frames from the pipe to TF. We as application developers could take advantage of TF's fast multi-threaded readers with minimal development effort. The function reads one frame from the FFmpeg buffer at a time, extracts the frame's timestamp and stores it in an array to be returned together with probabilities at the end of processing, then crops away extraneous edge pixels to maximize the sizes of image features after rescaling to the CNN's fixed input image size, and finally yields the cropped image to TF. The TF pipeline then applies model-specific preprocessing transformations to multiple frames concurrently, batches a specified number of frames together to be processed concurrently on the GPU, and lastly pre-loads the batch into GPU memory where it waits to start being processed immediately after the preceding batch finishes.

5.3.2 Model- and Hardware-Specific Video Frame Batching

Determining the optimal batch size for a given hardware configuration is a task left to the SNVA's users. If too low, the GPU risks being under-utilized. If too high performance may suffer, perhaps because there is insufficient available memory to pre-fetch the entire batch of images. For example, we found in our experiments with MobilenetV2 that a batch size of 64 was optimal in terms of frames processed per second even though our development machine's GPU easily supported a batch size of 128.

5.3.3 Using TensorRT and the Channels-First Data Format

Following TF and NVIDIA guidelines, we converted SNVA's models from the channels-last (NHWC) to the channels-first (NCHW) format in order to maximize inference performance. While we could not find and explicit explanation for the benefit, the use of NCHW is promoted in at least three locations in each of TF's and NVIDIA's documentation. Further, because SNVA takes advantage of TensorRT, and TensorRT only supported the NCHW format at the time of this writing, there was necessarily a transpose operation performed "under the hood" on each batch of images during processing. To avoid performing this computation on the GPU, and to gain the aforementioned unexplained benefits, we converted SNVA's models from NHWC to NCHW and added a transpose operation to the preprocessing pipeline that runs on CPU. We observed a 1% speed increase following the change, which added to the 10% speed increase gained by utilizing TensorRT in the first place.

5.3.4 Freezing TF Inference Graphs

Here, again, we follow guidance from the TF documentation and convert the model files output in TF checkpoint format during training, which contain large TF variable objects, to protobuf files in which all variables have been converted into constant values. As expected and intended, the performance gains are non-trivial.

5.4 Software Development Environment

SNVA was developed and alpha tested using an Alienware Area 51 workstation with a 10-core 3.00 GHz Intel Core i7-6950X CPU with hyperthreading, 64 GB of DDR4 SDRAM, and two NVIDIA GeForce GTX 1080 Ti graphics cards with 11GB of GDDR5 RAM each. The machine ran the Ubuntu 16.04 operating system, with NVIDIA driver 396.24, CUDA 9.0, cuDNN 7.0, TensorRT 3.0.4 , TensorFlow 1.8.0, Docker 18.03.1-CE, and NVIDIA-Docker 2.0.3 installed. The Python and Numpy versions used were 3.5 and 1.14, respectively.

When this setup was tested against all 31,535,862 video frames spanning 1,344 videos in the training pool, InceptionV3 inferred class labels at 826.24 fps on average over 10:40:04 h, MobilenetV2 (with one video processor assigned to each GPU) averaged 1491.36 fps over 05:58:20 h, and MobilenetV2 (with two video processors assigned to each GPU) averaged 1833.1 fps over 4:53:42 h. RAM consumption on our development machine appeared to be safely bounded above by 3.75 GB per active video analyzer.

5.5 Production SNVA Environment

The production environment to which SNVA was deployed included Dell PowerEdge C4130 servers featuring four NVIDIA Tesla V100 GPUs each. Given the use of Docker to deploy and run SNVA, the software environment in production is identical to that of development, with a marginal exception to the exact NVIDIA driver version. Ubuntu is also equal in version by construction. The videos are streamed from an NFS network share over a 16 Gb/s link. The fastest SNVA configuration (four MobilenetV2 analyzers across two GPUs) consumed videos at an average of 3.5 Gb/s when reading from an internal 7200 RPM hard disk drive. The first round of beta testing revealed that the V100 GPUs were not fully saturated by the processing pipeline developed using the 1080Ti GPUs, and that the application was actually CPU-bound. In response, the team intends to use cloud services to replicate the production environment closely and optimize the processing pipeline, potentially taking advantage of additional distributed computing resources available at VTTI.

6 Future Work

6.1 Precision-Oriented Active Learning

As was mentioned in Sect. 4, work zone scenes represent a small percentage of the total data. It may, therefore, be feasible to include samples misclassified as work zone scenes with high confidence in the labeling process once the model becomes sufficiently accurate. This means that in addition to model confidence on as yet unlabeled samples, and to the F0.5 score on the in-sample test set, precision on unlabeled samples can eventually become a candidate for measuring model performance. Recall that precision is the ratio of samples correctly classified as positive to all samples classified as positive. This would allow the unlabeled data to better indicate the model's potential efficacy on the target data set. Recall would be expected to improve though it would not be measured.

6.2 Robust Multi-frame Event Detection Using Bidirectional Recurrent Neural Networks

An early design of the SNVA application included a combination of a CNN as a work zone feature detector and a bidirectional RNN [22] as a work zone event detector, where an event may span many frames and include intermediate frames that may not actually contain work zone features. Consider for example a research volunteer driving through a work zone in the center lane when there is heavy traffic, leading to periodic occlusions of equipment by other vehicles. While the SHRP2 video data set, being static and historical rather than real-time and online is an attractive candidate for the application of an RNN that incorporates information from preceding and following frames into its prediction for the current frame, we ruled out the use of such an architecture for two reasons. First, we anticipated that budgetary and time constraints would not support the level of effort required to perform annotation at the video level. We turned out to be right. Second, after observing the number of frames that had to be excluded from consideration when developing the training set due to anomalies and ambiguities, it was not clear how video level annotation should be approached. As a research exercise not necessarily tied to SNVA, we intend to explore the feasibility of applying an RNN to SHRP2 video.

6.3 Other High-Priority Target Scene Features

As mentioned in the introduction, there exist a number of environmental conditions and roadway features that are not currently included in the RID but that researchers and other stakeholders have wanted to include in their studies dating back to the creation of the RID. Among them are traffic signal states and weather conditions. The SNVA application and the model development methodology presented here can readily be extended to support the detection of these features, and we intend to target them in future versions of SNVA.

7 Conclusion

We have introduced SNVA, the SHRP2 NDS Video Analytics software application. SNVA adds to the Roadway Information Database the ability for transportation safety researchers to formulate and ask questions specific to construction zones transited by Naturalistic Driving Study volunteer participant drivers. While the application was still undergoing beta testing at the time of this writing, alpha testing in the development environment implied that RID query results would be accurate and exhaustive. We described the approaches followed in developing the application as well as the machine learning model used to infer scene classifications from images in detail. The motivations behind the project and potential benefits to both data science and transportation communities if successful were also discussed. Currently, SNVA targets only one type of information: work zone presence, but in the near future, we intend to expand its capabilities to apply to weather events and traffic signal state. There are also opportunities to improve the application's robustness to fluctuations in the presence of features within an event region by feeding CNN feature vectors into a bidirectional recurrent neural network (likely an LSTM). The source code repository is publicly available at: https://github.com/VolpeUSDOT/SNVA

Acknowledgements Mr. Abodo gratefully acknowledges David Kuehn, Charles Fay, and Yusuf Mohamedshah of FHWA's Turner-Fairbank Highway Research Center (TFHRC), Miguel Perez, Joel Anderson and Calvin Winkowski of VTTI, Thomas Karnowski of Oak Ridge National Laboratory (ORNL) and Omar Smadi of Iowa State University's Center for Transportation Research and Education (CTRE) for their guidance and support during the development of SNVA. He would also like to thank his research advisor Leonardo Bobadilla and fellow Motion, Robotics and Automation (MoRA) lab mates Md. Mahbubur Rahman, Tauhidul Alam and Sebastián Zanlongo for providing instructive opportunities to engage in computer science research activities as an undergraduate at Florida International University's School of Computing and Information Sciences. This work was funded under Inter-Agency Agreement HW53A100 between the Volpe Center and FHWA.

References

1. S. Hallmark, O. Smadi, A. Sharma, Center for Transportation Research and Education at Iowa State University's Institute for Transportation (2015), https://shrp2.transportation.org/Documents/Safety/03-SHRP2%20IAP%20Round%204-MN-Work%20Zones%20v2.pdf
2. L. Cobb, Exploratory advanced research program - video analytics research projects. Technical report, 6300 Georgetown Pike, McLean, VA 22101 United States (2017)
3. T. Karnowski, R. Ferrell, Video reduction and analysis methods: Part ii, automated video analysis (2017), https://insight.shrp2nds.us/projectBackground/index
4. P. Viola, M. Jones, vol. 1 (2001), pp. I–511. https://doi.org/10.1109/CVPR.2005.177
5. N. Dalal, B. Triggs, in *2005 IEEE Computer Society Conference on Computer Vision and Pattern Recognition (CVPR'05)*, vol. 1 (2005), pp. 886–893. https://doi.org/10.1109/CVPR.2005.177
6. A. Krizhevsky, I. Sutskever, G.E. Hinton, Commun. ACM **60**, 84 (2012)
7. J. Paone, D. Bolme, R. Ferrell, D. Aykac, T. Karnowski, in *2015 IEEE Intelligent Vehicles Symposium (IV)* (2015), pp. 174–179. https://doi.org/10.1109/IVS.2015.7225682
8. M. Abadi, A. Agarwal, P. Barham, E. Brevdo, Z. Chen, C. Citro, G.S. Corrado, A. Davis, J. Dean, M. Devin, S. Ghemawat, I.J. Goodfellow, A. Harp, G. Irving, M. Isard, Y. Jia, R. Józefowicz, L. Kaiser, M. Kudlur, J. Levenberg, D. Mané, R. Monga, S. Moore, D.G. Murray, C. Olah, M. Schuster, J. Shlens, B. Steiner, I. Sutskever, K. Talwar, P.A. Tucker, V. Vanhoucke, V. Vasudevan, F.B. Viégas, O. Vinyals, P. Warden, M. Wattenberg, M. Wicke, Y. Yu, X. Zheng (2016), arXiv:1603.04467
9. O. Russakovsky, J. Deng, H. Su, J. Krause, S. Satheesh, S. Ma, Z. Huang, A. Karpathy, A. Khosla, M.S. Bernstein, A.C. Berg, F. Li (2014), arXiv:1409.0575
10. J. Yosinski, J. Clune, Y. Bengio, H. Lipson (2014), arXiv:1411.1792
11. C. Szegedy, V. Vanhoucke, S. Ioffe, J. Shlens, Z. Wojna (2015), arXiv:1512.00567
12. C. Szegedy, S. Ioffe, V. Vanhoucke (2016), arXiv:1602.07261
13. A.G. Howard, M. Zhu, B. Chen, D. Kalenichenko, W. Wang, T. Weyand, M. Andreetto, H. Adam (2017), arXiv:1704.04861
14. M. Sandler, A.G. Howard, M. Zhu, A. Zhmoginov, L. Chen (2018), arXiv:1801.04381
15. B. Zoph, V. Vasudevan, J. Shlens, Q.V. Le (2017), arXiv:1707.07012
16. K. He, X. Zhang, S. Ren, J. Sun (2016), arXiv:1603.05027
17. K. Simonyan, A. Zisserman (2014), arXiv:1409.1556
18. R.A. Baeza-Yates, B. Ribeiro-Neto, *Modern Information Retrieval* (Addison-Wesley Longman Publishing Co. Inc, Boston, 1999)
19. T. Ching, D.S. Himmelstein, B.K. Beaulieu-Jones, A.A. Kalinin, B.T. Do, G.P. Way, E. Ferrero, P.M. Agapow, M. Zietz, M.M. Hoffman, W. Xie, G.L. Rosen, B.J. Lengerich, J. Israeli, J. Lanchantin, S. Woloszynek, A.E. Carpenter, A. Shrikumar, J. Xu, E.M. Cofer, C.A. Lavender, S.C. Turaga, A.M. Alexandari, Z. Lu, D.J. Harris, D. DeCaprio, Y. Qi, A. Kundaje, Y. Peng, L.K. Wiley, M.H.S. Segler, S.M. Boca, S.J. Swamidass, A. Huang, A. Gitter, C.S.

Greene, J. Royal Soc. Interface **15**, 141 (2018). https://doi.org/10.1098/rsif.2017.0387, http://rsif.royalsocietypublishing.org/content/15/141/20170387
20. G. Inc. Tensorflow-slim image classification model library (2017), https://github.com/tensorflow/models/tree/master/research/slim
21. D. D.Lewis, J. Catlett, Eleventh International Conference on Machine Learning (1994), https://www.sciencedirect.com/science/article/pii/B978155860335650026X
22. A. Graves, A. Mohamed, G.E. Hinton (2013), arXiv:1303.5778

Franklin Abodo is a Master's student of computer science at Florida International University in Miami, FL, and a Student Trainee Computer Scientist at the U.S. DOT Volpe Center where he works as a software and machine learning engineer, and collaborates with city planners and transportation engineers on the advancement of traffic simulation models.

Robert Rittmuller is an Information Technology Specialist at the U.S. DOT Volpe Center, with additional roles as project manager and business developer for various machine learning and AI projects. He provides technical support and expertise to several U.S. DOT groups such as NHTSA, FMCSA, FHWA, and the Intelligent Transportation Systems Joint Program Office (ITS-JPO). Prior to joining Volpe, he worked in executive management for an internet startup. He received his undergraduate degree from Harvard University in 2008 and currently holds several industry certifications such as CISSP, PMP, and VCP. He also pursues his love of photography when he has time, and his photos have been featured on the History Channel. He currently resides in Salem, MA with his wife Karen and his son Zachary.

Brian Sumner received his bachelor's degree in Computer Science from Tufts University in 2013. He currently works as an IT Specialist at the U.S. DOT Volpe Center, where he develops software applications for a variety of U.S. DOT sponsor agencies.

Andrew Berthaume earned his Ph.D. in Civil Engineering from the University of Massachusetts Amherst in 2015. He currently manages several research projects at the U.S. DOT Volpe Center. Among his projects are a driver behavior modeling project with FHWA in which he uses naturalistic driving data (such as the SHRP2 NDS data) to study driver behaviors and trends under specific driving conditions, and to develop novel traffic models that specially consider those conditions.

Action Recognition in Videos Using Multi-stream Convolutional Neural Networks

Helena de Almeida Maia, Darwin Ttito Concha, Helio Pedrini, Hemerson Tacon, André de Souza Brito, Hugo de Lima Chaves, Marcelo Bernardes Vieira and Saulo Moraes Villela

Abstract Human action recognition aims to classify trimmed videos based on the action being performed by one or more agents. It can be applied to a large variety of tasks, such as surveillance systems, intelligent homes, health monitoring, and human-computer interaction. Despite the significant progress achieved through image-based deep networks, video understanding still faces challenges in modeling spatiotemporal relations. The inclusion of temporal information in the network may lead to significant growth in the training cost. To address this issue, we explore complementary handcrafted features to feed pre-trained two-dimensional (2D) networks in a multi-stream fashion. In addition to the commonly used RGB and optical flow streams, we propose the use of a stream based on visual rhythm images that encode long-term information. Previous works have shown that either RGB or optical flow streams may benefit from pre-training on ImageNet since they maintain a certain

H. de Almeida Maia (✉) · D. T. Concha · H. Pedrini
Institute of Computing, University of Campinas, Campinas, SP, Brazil
e-mail: helena.maia@liv.ic.unicamp.br

D. T. Concha
e-mail: darwin.ttito@liv.ic.unicamp.br

H. Pedrini
e-mail: helio@ic.unicamp.br

H. Tacon · A. de Souza Brito · H. de Lima Chaves · M. B. Vieira · S. M. Villela
Department of Computer Science, Federal University of Juiz de Fora, Juiz de Fora, MG, Brazil
e-mail: hemerson@ice.ufjf.br

A. de Souza Brito
e-mail: andre.brito@ice.ufjf.br

H. de Lima Chaves
e-mail: hugo.lima.chaves@ice.ufjf.br

M. B. Vieira
e-mail: marcelo.bernardes@ufjf.edu.br

S. M. Villela
e-mail: saulo.moraes@ufjf.edu.br

M. A. Wani et al. (eds.), *Deep Learning Applications*,
Advances in Intelligent Systems and Computing 1098,
https://doi.org/10.1007/978-981-15-1816-4_6

level of object shape. The visual rhythm, on the other hand, harshly deforms the silhouettes of the actors and objects. Therefore, we develop a different pre-training procedure for the latter stream using visual rhythm images extracted from a large and challenging video dataset, the Kinetics.

1 Introduction

In recent years, a large amount of video data has been produced and released due to the easy access to both the equipments for capturing new data such as video cameras and mobiles, and platforms such as YouTube for sharing. For this reason, many video datasets have become available, enabling the research and development of various applications based on video analysis in public, private, and restricted areas such as streets, banks, and radioactive places. Since the analysis of large amounts of data by human operators may be stressful and may involve sensitive content, automatic procedures are needed to address related problems.

The problem addressed in this work is the recognition of human actions in videos [1, 5, 15, 19, 29, 34, 43] that aims to detect and identify actions of one or more agents. It is a challenging task since the same action may vary according to the actor, and the scene may present difficult conditions, such as occlusions, background clutter, and camera motion. This problem has several relevant applications, such as intelligent surveillance [9] and human-computer interaction [7, 22].

The majority of current approaches that address this problem employ deep learning, since it has shown to be a useful tool to generalize data in complex scenarios, achieving impressive results in different computer vision problems (for instance, image classification). However, the inclusion of temporal information may increase the number of parameters in the network, leading to a significant increase in the training cost. Moreover, designing spatiotemporal models brings a major issue by choosing a proper temporal extension that encloses every possible action without compromising the computational cost. For this reason, many recent deep learning proposals have explored handcrafted inputs, such as optical flow images, in order to encode action dynamics. Image networks and fusion techniques are used to process these inputs and capture temporal evolution [10, 11, 16, 20, 24, 38, 40].

In this paper, we propose a three-stream architecture based on the two-stream one [24] that explores complementary modalities to recognize the action, RGB frames (spatial) and optical flow images (temporal). In our architecture, a third modality called visual rhythm is used to provide dynamic information of the entire video for the network. Spatial and temporal streams have achieved great results using the parameters from ImageNet training for initialization, but our third modality has a very different nature, requiring a proper pre-training. Therefore, an additional contribution of our work is the use of visual rhythms computed from the large Kinetics dataset to pre-train the third stream. Moreover, we study different fusion strategies to combine the outputs of the three streams.

Experiments conducted on two well-known challenging datasets, UCF101 (University of Central Florida [26]) and HMDB51 (Human Motion DataBase [14]), achieved accuracy rates comparable to state-of-the-art approaches, which demonstrates the effectiveness of the visual rhythm as a spatiotemporal feature.

This text is organized as follows: In Sect. 1, we introduced the problem of human action recognition in videos addressed in this work. State-of-the-art approaches and basic concepts are presented in Sects. 2 and 3, respectively. Further details about our proposed architecture are given in Sect. 4. Experimental results of the method are presented and discussed in Sect. 5. Finally, in Sect. 6, we present the conclusions of our work.

2 Related Work

Most of the first attempts to employ deep learning techniques in the human action recognition context were based on static information using image networks, due to the complexity of designing and training video networks from scratch. Since the temporal dimension is rather relevant to determine the action, static information from different frames is aggregated in a fusion stage [10, 11, 20]. However, the network responsible for extracting the features is not aware of the action dynamics.

A notable architecture was proposed by Simonyan and Zisserman [24] that uses two image networks in parallel: a static network that processes one RGB frame per video to capture the scene appearance and a dynamic one based on optical flow images to capture short-term motion (approximately 10 frames). The combination of these complementary modalities achieved promising results and gave rise to a variety of state-of-the-art approaches.

To consider longer temporal evolution, Ng et al. [16] repeated the feature extraction process for several frames and optical flow images on a given video. They considered two different feature aggregation methods: pooling layers and LSTM cells (long short-term memory). This architecture was capable of processing up to 120 frames per video. A similar approach was proposed by Wang et al. [38], the Two-Stream TSN (Temporal Segment Network). In the Two-Stream TSN, however, the predictions are fused, instead of the features, using a segmental consensus function that does not impose temporal limits. Wang et al. [39] proposed a hierarchical fusion strategy, which is also based on RGB and optical flow modalities. According to their experiments, the best network covers up to 30 frames per video. Wang et al. [40], in turn, used the dynamic image in a third and new stream, encoding simultaneously appearance and motion information along 20 consecutive frames.

Carreira and Zisserman [3] proposed a new network by inflating state-of-the-art image architectures into 3D CNNs. This method is also based on complementary information and is called Two-Stream Inflated 3D ConvNets (Two-Stream I3D). Both the filters and the parameters are replicated along the time dimension for the conversion into 3D. Besides the traditional ImageNet pre-training step, an additional one is performed using a large video dataset (approximately 300 k clips), the Kinetics [12]. This extra step leads to a considerable increase in the accuracies, especially on HMDB51 dataset.

In fact, the current best results on UCF101 and HMDB51 datasets are achieved by exploring the Kinetics to pre-train the network [3, 5, 33, 41, 45]. Choutas et al. [5] and Wang et al. [41] used the I3D model combined with their own methods. The DTPP (Deep networks with Temporal Pyramid Pooling [45]) is a method that aggregates frame-level features using multiple pooling layers to obtain a video-level representation. In order to perform spatiotemporal convolutions, Tran et al. [33] factorized 3D convolutional filters into spatial and temporal components (2 + 1D), achieving superior results when compared to 2D CNN. All of these works improved their initial results by pre-training their networks on Kinetics or using the I3D which was already pre-trained. However, due to the Kinetics size, this strategy requires further computational resources.

3 Basic Concepts

In this section, two important concepts related to the problem under investigation are briefly described.

3.1 *Visual Rhythm*

The visual rhythm [13] is a feature originated from the spatiotemporal slices [17] that encodes the entire video in a single image. The resulting image is obtained by the concatenation of subsampled pixels or a predefined 1D feature computed from each frame, which are called slices.

By choosing proper slices, the visual rhythm may contain rich information to detect and classify events in the video. For instance, Ngo et al. [17] proposed a method for locating and classifying video transitions through the analysis of the central row (horizontal), central column (vertical), and the main diagonal subsampled from the frames. Torres and Pedrini [32] explored these 2D images to estimate object trajectories throughout the video in a handcrafted process applied to three computer vision problems: abnormal event detection, human action recognition, and hand gesture recognition.

Formally, let $V = \{F_1, F_2, \ldots, F_t\}$ be a video with t frames F_i, where each frame is an $h \times w$ matrix. Consider $T(F_i) = S_i$ an operation that maps a frame F_i into an $n \times 1$ column vector S_i, either by subsampling or computing features from F_i. The visual rhythm for the entire video V is given by the $n \times t$ matrix:

$$\mathrm{VR}(V) = [T(F_1)\; T(F_2)\; \cdots\; T(F_t)] = [S_1\; S_2\; \cdots\; S_t]. \quad (1)$$

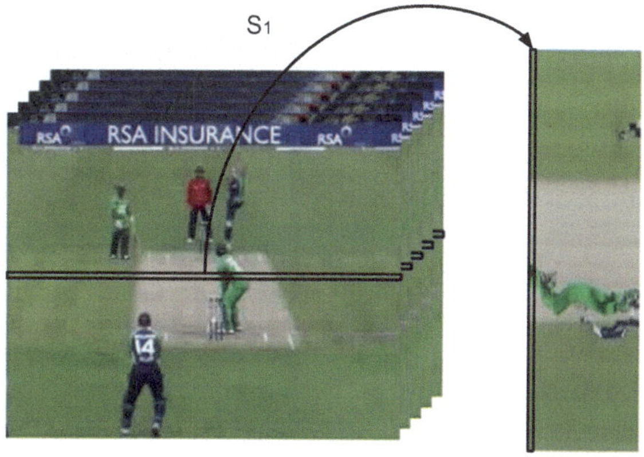

Fig. 1 Example of visual rhythm image generated for CricketShot class from the UCF101 dataset. The central row of each frame becomes a slice in the resulting image

Figure 1 shows an example of visual rhythm construction, where each slice corresponds to the central row of a frame and is placed in a column of the resulting image. Considering the video as a volume XYT, the resulting image can be seen as a plane parallel to XT.

Here, we use the operation proposed by Souza [27] to obtain two different visual rhythms: horizontal-mean and vertical-mean. Consider $F_i(y, x)$ the RGB value of the frame F_i in the (y, x) coordinates, with $y \in \{1, \ldots, h\}$ and $x \in \{1, \ldots, w\}$, the operations $T_h(F_i)$ and $T_v(F_i)$ are given by the mean intensity of the columns (horizontal-mean) or rows (vertical-mean) of F_i. That is, the horizontal-mean visual rhythm is defined as

$$T_h(F_i) = \left[\frac{\sum_y F_i(y, 1)}{h} \; \frac{\sum_y F_i(y, 2)}{h} \; \cdots \; \frac{\sum_y F_i(y, w)}{h} \right]^T \tag{2}$$

and the vertical-mean one as:

$$T_v(F_i) = \left[\frac{\sum_x F_i(1, x)}{w} \; \frac{\sum_x F_i(2, x)}{w} \; \cdots \; \frac{\sum_x F_i(h, x)}{w} \right]^T. \tag{3}$$

Both directions are illustrated in Fig. 2. Note that, unlike Fig. 1, every pixel from a given frame is taken into account to compute the corresponding slice using either of these two operations.

3.2 Two-Stream Architecture

In the two-stream network [24], a single RGB frame is randomly selected from the video, as well as 10 pairs of consecutive optical flow images in the form of a 20-channel image to respectively train the spatial and temporal streams.

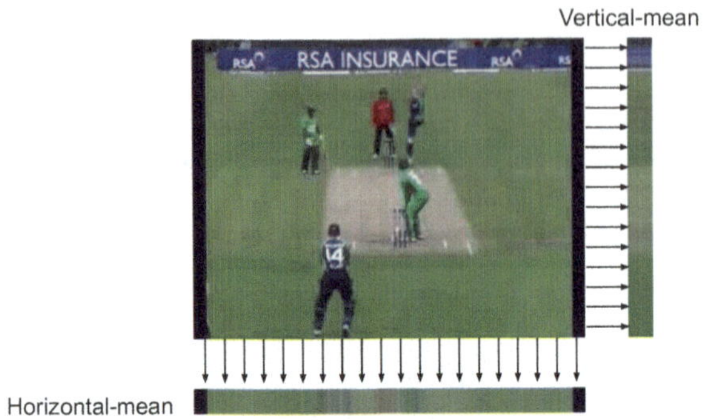

Fig. 2 Spatiotemporal slice: the horizontal-mean/vertical-mean slice from a given frame contains the average of the columns/rows. The slices were resized for illustration purposes

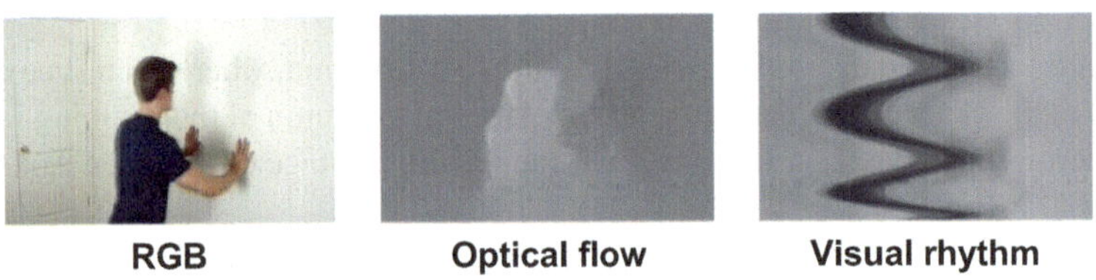

Fig. 3 Modalities used in our three-stream network. The object silhouette can be seen in the RGB and optical flow images, but not in the visual rhythm image

Although dynamic information is rather relevant for action recognition, static and context information such as actor poses, the objects involved and standard scenarios may help distinguish the classes. A green grass field, for instance, maybe a clue for actions related to soccer games; a horse may help to recognize the horse riding action. For this reason, even using a single video frame, the spatial stream alone is capable of achieving good results.

Each stream is individually trained. Since the spatial stream works with the same modality as image networks for classification and has a comparable goal (appearance recognition), it is reasonable that it can be pre-trained using image datasets such as ImageNet [21], followed by fine-tuning on the desired video dataset. Surprisingly, experiments indicate that the same pre-training process may be applied to the temporal stream [37]. We believe that this knowledge transferring is possible because the optical flow maintains a certain level of object shape, especially when compared to visual rhythm images (Fig. 3).

The original network is based on CNN-M-2048 [4], which is composed of five convolutional layers followed by three fully connected ones. However, Wang et al. [37] argued that deeper networks, such as VGG [25] (16 or 19 layers) and GoogLeNet [30] (22 layers), are preferable to address our target problem, since the concept of action is more complex than the object. Wani et al. [42] discuss several CNN architectures with a variety of depths. Among the deepest CNNs, we tested the

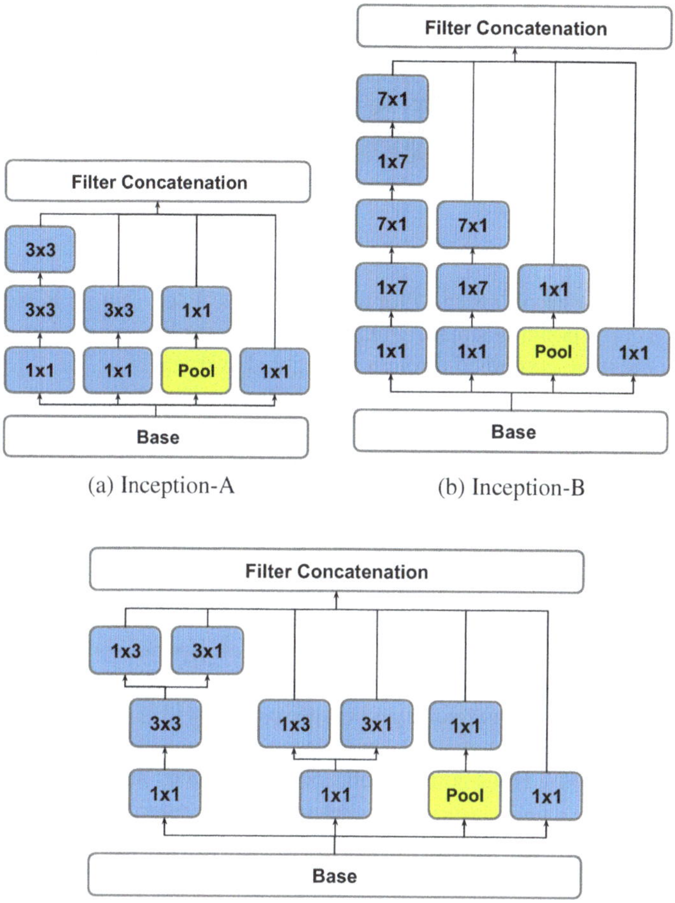

(a) Inception-A (b) Inception-B

(c) Inception-C

Fig. 4 Inception modules. Blue and yellow rectangles represent convolutional and pooling layers, respectively

ResNet [8] (152 layers) and the Inception V3 [31] (48 layers) in our previous work [6]. The Inception V3 presented slightly better accuracies and so we have chosen it to compound our multi-stream network. Similar to the GoogleNet (also known as Inception V1), the Inception V3 is based on inception modules (Fig. 4). The authors explored factorized convolutions to build an efficient network with a low number of parameters. The Inception V3 architecture is shown in Fig. 5.

For the spatial and temporal streams, the training data is augmented using random cropping, horizontal flipping, and RGB jittering. To avoid overfitting in very deep CNNs, two additional data augmentation techniques were proposed by Wang et al. [37]: corner and multiscale cropping. For testing, 25 frames/stacks of optical flow images are selected from each video and used to produce 10 new samples

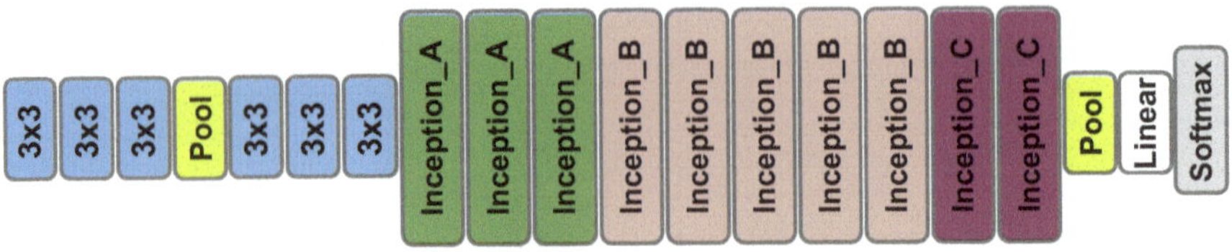

Fig. 5 Inception V3 architecture

per frame/stack by cropping and flipping techniques. Each sample is individually tested in the corresponding stream. Finally, the class scores computed in each CNN (softmax scores) are combined through a weighted average.

4 Proposed Method

An overview of our three-stream network is shown in Fig. 6. It contains three deep CNNs working with different modalities: RGB frame (spatial), optical flow (temporal), and visual rhythm (spatiotemporal). Each one is an image network pre-trained in ImageNet and independently fine-tuned with its corresponding modality. The spatiotemporal stream has an extra pre-training step using the Kinetics. All the training data is augmented using multiscale and corner cropping [37], and random horizontal flipping. During the test stage, 10 samples are produced from each input image using corner cropping (four corners and one central crop) and horizontal flipping techniques.

Given a video, the output of each stream is a feature vector containing the softmax scores for the m classes. A fusion strategy is applied to obtain a single score vector for the input video. Further details about the streams are given as follows:

4.1 Improved Spatial Stream

In our improved spatial stream, instead of collecting a single frame per video, we randomly collect two frames, one in each half of the video. This approach is justified by the fact that the appearance of the scene may change significantly during the time, either by scene conditions such as lighting and occlusions or by the variety of poses, objects, and background in the video. The CNN still receives one of those frames at a time in the training stage. However, by presenting two samples taken at different positions of the video, we are able to capture variations in appearance such as different backgrounds that may be characteristics of certain actions.

The testing protocol remains the same in our spatial stream, we use 25 frames evenly sampled from each testing video and 10 new samples are produced from

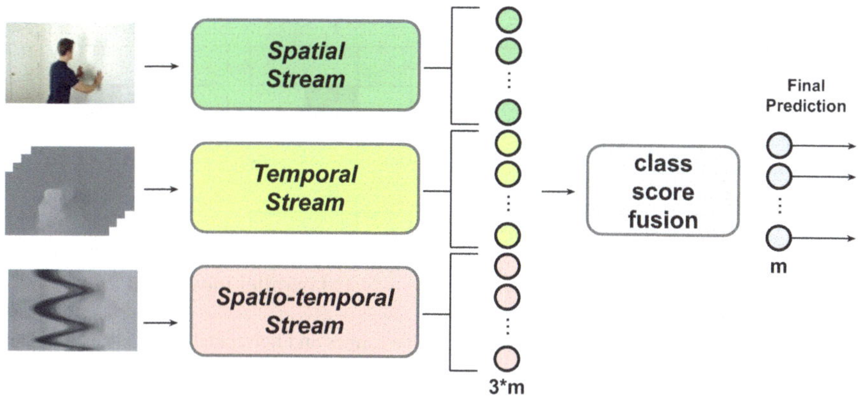

Fig. 6 Overview of our three-stream proposal for action recognition

them. Each sample is individually tested, and all the computed outputs are combined through the average of the scores to obtain the stream m-dimensional vector.

4.2 *Temporal Stream*

The temporal stream is trained with 10 pairs of consecutive optical flow images per video, in the form of a 20-channel image (stack). Each pair of optical flow images represents the motion along X and Y axes between two consecutive frames, and the whole stack encodes short-term information (10 frames) about the dynamics of the action.

For testing, 25 stacks of optical flow images are evenly sampled from each video and used to produce 10 new samples per stack. Similar to the spatial stream, the 250 temporal outputs for a given video are combined using the average of the scores, generating a single m-dimensional vector per video.

4.3 *Spatiotemporal Stream*

The spatiotemporal stream receives as input a single visual rhythm extracted from the video. The visual rhythm is a grayscale image computed with the Eq. 2 (horizontal-mean) or Eq. 3 (vertical-mean).

We propose a method for adaptively deciding the best visual rhythm direction for each action according to the predominant movement. The method is called Adaptive Visual Rhythm (AVR) and is based on the following observation. Consider a fixed column/row j, the set $\{S_1(j), S_2(j), \ldots, S_t(j)\}$ represents the variation in the average

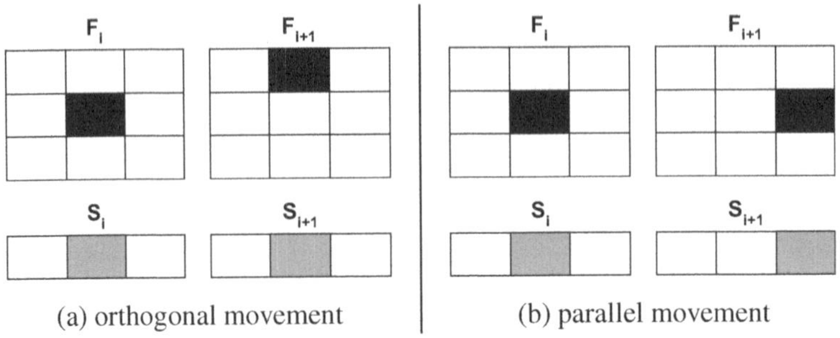

Fig. 7 A moving object considering two consecutive frames and horizontal-mean slices. Parallel movement is better captured in the slice

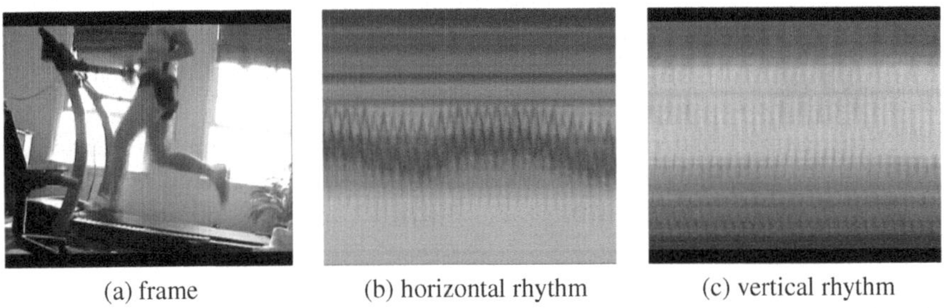

Fig. 8 Example of frame and rhythms from a Kinetics video of the "running on treadmill" class. The horizontal rhythm presents a wavy pattern that better characterizes the action

value regarding j across the time and can be seen in the jth row of the horizontal-mean/vertical-mean rhythm. If the mean value remains constant for the column/row j, the jth row in the rhythm will form a line with homogeneous intensity.

Suppose, without loss of generality, that we are working with horizontal-mean slices. If a given object moves vertically (i.e. orthogonally to the slice direction) between two frames, it is very likely that the mean color of the corresponding column remains the same (Fig. 7). However, a horizontal movement affects the average color of all columns spanned by the object. Therefore, movements parallel to the slice direction tends to produce more distinctive patterns.

For estimating the predominant direction of movement, we use the Lucas–Kanade point-tracker [2]. The method tracks a set of selected points [23] across the video. For every video of a given class, the absolute horizontal and vertical displacements estimated by the tracker are accumulated over the frames, resulting in two values per class. The highest value defines the class rhythm direction. That is, we choose the horizontal-mean rhythm if the horizontal movement is predominant in the class and the vertical-mean otherwise. This process is only performed once, and the determined directions are used for any posterior training.

Figure 8 shows a frame and the visual rhythms extracted from a Kinetics video of the "running on treadmill" class. This action is predominantly horizontal due to the leg motion. For this reason, the horizontal rhythm presents more relevant patterns

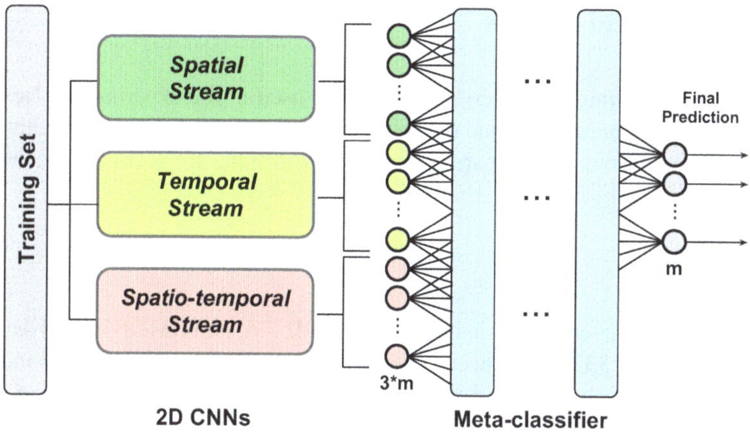

Fig. 9 Stacking: the vectors computed by the streams are fused using fully connected layers to obtain the final prediction for each video. The sizes of the input and output are based on the number of classes in the dataset

for the classification. As can be observed in the example, the horizontal rhythm contains a wavy pattern that represents the leg movements, whereas the vertical one is composed of quite homogeneous lines.

4.4 Stacking

In our experiments, we noticed that an improvement in the individual streams does not necessarily imply an improvement in the combination. For this reason, a good fusion strategy is fundamental for the method's effectiveness.

In this work, besides the simple and weighted average fusion [6, 24, 37], we explore another fusion strategy using external fully connected (FC) layers as a meta-classifier. Thus, the network automatically defines how much the features contribute to the final prediction. The external network is trained using the same training set as the 2D CNNs, but using the features computed by them as input (Fig. 9). This combination of classifiers is called stacking.

The training procedure is divided into two stages: (1) 2D CNNs and (2) meta-classifier training. The input of the meta-classifier is formed by the concatenation of the stream outputs. The output is an m-dimensional vector with the class scores.

The idea behind this proposal is to learn misclassification patterns from the streams combined with the others. That is, if two classes are poorly discriminated in a given stream, but well classified in another one, the external network may capture this pattern. The main advantage of this method is the automatic weight assignment that adapts to the inclusion of new streams and modifications in the approaches.

5 Experimental Results

In this section, we initially describe the datasets used in the experiments. Next, we present the setting parameters and the results obtained with the proposed method. At the end of the section, we compare our method with state-of-the-art approaches.

5.1 *Datasets*

Our method is evaluated on the challenging UCF101 [26] and HMDB51 [14] datasets. UCF101 contains 13320 sequences collected from YouTube and is divided into 101 classes. The samples have a fixed resolution of 320×240 pixels, frame rate of 25 fps and various lengths. The dataset also includes recommended splits into approximately 70-30 for training and testing, respectively. HMDB51 is composed of 6766 sequences extracted from various sources, mostly from movies. The samples are categorized into 51 action classes. Since it combines commercial and non-commercial sources, it presents a rich variety of sequences, including blurred videos or with lower quality and actions from different point of view. The authors also provide three recommended splits, where each one contains 70 samples for training and 30 for testing per action class. The overall performance in both datasets is assessed by the average classification accuracy achieved in the splits.

For the pre-training of the spatiotemporal stream, we use the Kinetics [12] dataset, which contains 600 action classes with 600–1150 clips per class and approximately 10 s per clip. Kinetics is a large dataset that has a total of 495547 clips taken from YouTube. The number of clips in the recommended training/validation/testing set is 450–1000, 50 and 100 clips per class, respectively.

5.2 *Results*

The baseline for our work was provided by Zhu [46], which is a public implementation of the very deep two-stream network [37] using PyTorch framework. The Inception V3 [31] architecture was adopted as CNN for our three streams using the ImageNet parameters for initialization provided by PyTorch. All experiments were performed on a machine with an Intel® Core™ i7-3770K 3.50 GHz processor, 32 GB of memory, an NVIDIA GeForce® GTX 1080 GPU and Ubuntu 16.04.

After the ImageNet initialization, the spatiotemporal stream is trained using the Kinetics. For both, pre-training on Kinetics and training on UCF101/HMDB51, we use 250 epochs and a learning rate of 0.001. Since the Kinetics contains a validation set, we use it to determine the best pre-trained model for the posterior UCF101/HMDB51 training. According to our experiments, the best pre-trained model was the one at the 50th epoch. The UCF101 and HMDB51, on the other hand, do not provide a validation set, and so we use the model of the last epoch for testing.

Table 1 Spatiotemporal stream results using different pre-training datasets

Pre-training dataset	UCF101 (%)	HMDB51 (%)
ImageNet [6]	64.74	39.63
ImageNet + Kinetics	**66.68**	**48.91**

Table 2 Different strategies for fusing the outputs of the three streams

Approach	Streams	UCF101 (%)	HMDB51 (%)
Individual	RGB* images [6]	86.61	51.77
	Optical flow [6]	86.95	59.91
	AVR-K	66.68	48.91
Simple average	RGB* + Optical flow	93.08	65.03
	RGB* + AVR-K	87.37	62.00
	Optical flow + AVR-K	82.66	61.55
	RGB* + Optical flow + AVR-K	91.97	67.65
Weighted average	RGB* + Optical flow	93.06	65.80
	RGB* + AVR-K	90.73	63.12
	Optical flow + AVR-K	88.52	66.43
	RGB* + Optical flow + AVR-K	**93.91**	**70.07**
Stacking	RGB* + Optical flow	92.35	65.05
	RGB* + AVR-K	88.58	61.70
	Optical flow + AVR-K	85.79	60.13
	RGB* + Optical flow + AVR-K	91.94	67.14

Results with and without the pre-training on Kinetics are shown in Table 1. The extra pre-training step improves the results in both datasets, specially HMDB51 that has an increase of 9.28%. This may be explained by the size of the datasets since HMDB51 is half the size of UCF101 and is more challenging, it takes greater benefit from the pre-training.

In Table 2, we show the results for each of the three streams separately and every combination of them using different fusion strategies. AVR-K refers to the AVR pre-trained with ImageNet and Kinetics. In both datasets, UCF101 and HMDB51, the temporal stream achieves the best individual results, as in the original two-stream network. For the combination, we consider the simple and weighted averages of the output vectors, and stacking with a meta-classifier. In the weighted average, we use the weight two for the spatial, three for the temporal, and one for the spatiotemporal stream, based on the individual performances. Our meta-classifier is composed of a single FC layer that receives the concatenation of the stream vectors as input and returns a vector with the scores for each class.

Concerning the two-stream combinations, most of them surpass the individual results, except for optical flow + AVR-K with simple average and stacking on the UCF101. In these two cases, the combinations perform worse than optical flow alone.

Table 3 Comparison of accuracy rates (%) for UCF101 and HMDB51 datasets. Cells on bold represents the overall highest accuracy rates, whereas underlined cells consist of the best results using only ImageNet to pre-train the network

Method	Pre-training dataset	UCF101 (%)	HMDB51 (%)
iDT + HSV [18]	—	87.9	61.1
Two-stream [24]	ImageNet	88.0	59.4
Two-stream + LSTM [16]	ImageNet	88.6	—
Two-stream TSN [38]	ImageNet	94.0	68.5
Three-stream TSN [38]	ImageNet	94.2	69.4
Three-stream [40]	ImageNet	94.1	<u>70.4</u>
TDD + iDT [36]	ImageNet	91.5	65.9
LTC + iDT [35]	—	92.7	67.2
KVMDF [44]	ImageNet	93.1	63.3
STP [39]	ImageNet	<u>94.6</u>	68.9
L^2STM [28]	ImageNet	93.6	66.2
Two-stream + AVR (ResNet 152) [6]	ImageNet	94.3	68.3
Two-stream + AVR (Inception V3) [6]	ImageNet	93.7	69.9
Two-stream I3D [3]	ImageNet + Kinetics	98.0	80.9
I3D + PoTion [5]	ImageNet + Kinetics	**98.2**	80.9
DTPP [45]	ImageNet + Kinetics	98.0	**82.1**
SVMP + I3D [41]	ImageNet + Kinetics	—	81.3
R(2 + 1)D-TwoStream [33]	Kinetics	97.3	78.7
Our method (weighted average)	ImageNet + Kinetics	93.9	70.1

In general, the presence of the spatial stream boosts the accuracies in comparison with optical flow + AVR-K, which suggests the importance of appearance information for the recognition.

For the HMDB51, the three-stream versions outperform all other combinations, whereas for the UCF101 this can only be verified for the weighted average fusion. However, the gain in the challenging HMDB51 (approximately +3%) is more significant than the loss in the UCF101 (approximately −1%). We can conclude from this table that the weighted average of the three outputs outperform the other fusion strategies, perhaps because it combines the scores of the same class instead of the interclass combination of the FC layer, and assigns better weights than the simple average.

After showing and analyzing the results obtained in both datasets, we compare our accuracies with the state-of-the-art approaches presented in Table 3. We separate the methods according to the pre-training strategy. Some of them are not based on deep networks and so they do not have a pre-training step. Concerning the remaining methods, one group pre-train on ImageNet dataset, whereas the other uses the ImageNet and Kinetics. The works pre-trained with both have an advantage over those trained only with ImageNet, since the Kinetics is one of the largest and most varied dataset for action recognition problem. Moreover, some of them, such as the I3D, is based on 3D convolutions. Our method achieves competitive results, considering the first group. However, we can see that the gain in the individual results compared to the original AVR (Table 1) is not reflected in the fusion. The accuracy rates of our three-stream network are only 0.2% greater than the original two-stream + AVR using the Inception V3, despite the increase of 1.94% (UCF101) and 9.28% (HMDB51) in the individual results. Therefore, further analysis on the fusion strategy is required to improve the results.

6 Conclusions

In this work, we proposed an innovative approach to the problem of human action recognition in video sequences. We explored information from spatial, temporal, and spatiotemporal networks based on a multi-stream architecture. Since the visual rhythms from the spatiotemporal stream deformed significantly the object silhouettes, this stream required an additional pre-training step using the same modality instead of natural images.

We combined the individual outputs computed in each stream by means of three different fusion techniques: simple average, weighted average, and FC layer. The weighted average achieved the best results on the UCF101 and HMDB51 datasets. Our method achieved promising results compared to state-of-the-art approaches.

Acknowledgements The authors thank FAPESP (grants #2017/09160-1 and #2017/12646-3), CNPq (grant #305169/2015-7), CAPES and FAPEMIG for their financial support. The authors are also grateful to NVIDIA for the donation of a GPU as part of the GPU Grant Program.

References

1. H. Bilen, B. Fernando, E. Gavves, A. Vedaldi, Action recognition with dynamic image net works. IEEE Trans. Pattern Anal. Mach. Intell. **40**(12), 2799–2813 (2018)
2. J.Y. Bouguet, Pyramidal implementation of the affine lucas kanade feature tracker description of the algorithm. Intel Corp. **5**(1–10), 4 (2001)
3. J. Carreira, A. Zisserman, Quo vadis, action recognition? a new model and the kinetics dataset, in *IEEE Conference on Computer Vision and Pattern Recognition* (IEEE, 2017), pp. 4724–4733
4. K. Chatfield, K. Simonyan, A. Vedaldi, A. Zisserman, Return of the devil in the details: delving deep into convolutional nets (2014), pp. 1–11, arXiv:14053531
5. V. Choutas, P. Weinzaepfel, J. Revaud, C. Schmid, PoTion: pose motion representation for action recognition, in *IEEE Conference on Computer Vision and Pattern Recognition* (2018), pp. 7024–7033
6. D.T. Concha, H. de Almeida Maia, H. Pedrini, H. Tacon, A. de Souza Brito, H. de Lima Chaves, M.B. Vieira Multi-stream convolutional neural networks for action recognition in video sequences based on adaptive visual rhythms, in *IEEE International Conference on Machine Learning and Applications* (IEEE, 2018), pp. 473–480

7. I. Gori, J.K. Aggarwal, L. Matthies, M.S. Ryoo, Multitype activity recognition in robot-centric scenarios. IEEE Robot. Autom. Lett. **1**(1), 593–600 (2016)
8. K. He, X. Zhang, S. Ren, J. Sun, Deep residual learning for image recognition, in *IEEE Conference on Computer Vision and Pattern Recognition* (2016), pp. 770–778
9. S. Ji, W. Xu, M. Yang, K. Yu, 3D convolutional neural networks for human action recognition. IEEE Trans. Pattern Anal. Mach. Intell. **35**(1), 221–231 (2013)
10. R. Kahani, A. Talebpour, A. Mahmoudi-Aznaveh, A correlation based feature representation for first-person activity recognition. Multimed. Tools Appl. **78**(15), 21673–21694 (2019)
11. A. Karpathy, G. Toderici, S. Shetty, T. Leung, R. Sukthankar, L. Fei-Fei, Large-scale video classification with convolutional neural networks, in *IEEE Conference on Computer Vision and Pattern Recognition* (2014), pp. 1725–1732
12. W. Kay, J. Carreira, K. Simonyan, B. Zhang, C. Hillier, S. Vijayanarasimhan, F. Viola, T. Green, T. Back, P. Natsev, M. Suleyman, A. Zisserman, The kinetics human action video dataset (2017), pp. 1–22, arXiv:170506950
13. H. Kim, J. Lee, J.H. Yang, S. Sull, W.M. Kim, S.M.H. Song, Visual rhythm and shot verification. Multimed. Tools Appl. **15**(3), 227–245 (2001)
14. H. Kuehne, H. Jhuang, R. Stiefelhagen, T. Serre, HMDB51: a large video database for human motion recognition, *High Performance Computing in Science and Engineering* (Springer, Berlin, 2013), pp. 571–582
15. D. Li, T. Yao, L. Duan, T. Mei, Y. Rui, Unified spatio-temporal attention networks for action recognition in videos. IEEE Trans. Multimed. 416–428 (2018)
16. J.Y.H. Ng, M. Hausknecht, S. Vijayanarasimhan, O. Vinyals, R. Monga, G. Toderici, Beyond short snippets: deep networks for video classification, in *IEEE Conference on Computer Vision and Pattern Recognition* (2015), pp. 4694–4702
17. C.W. Ngo, T.C. Pong, R.T. Chin, Camera break detection by partitioning of 2D spatio-temporal images in MPEG domain, in *IEEE International Conference on Multimedia Computing and Systems*, vol. 1 (IEEE, 1999), pp. 750–755
18. X. Peng, L. Wang, X. Wang, Y. Qiao, Bag of visual words and fusion methods for action recognition: comprehensive study and good practice. Comput. Vis. Image Underst. **150**, 109–125 (2016)
19. H. Rahmani, A. Mian, M. Shah, Learning a deep model for human action recognition from novel viewpoints. IEEE Trans. Pattern Anal. Mach. Intell. **40**(3), 667–681 (2018)
20. M. Ravanbakhsh, H. Mousavi, M. Rastegari, V. Murino, L.S. Davis, Action recognition with image based CNN features (2015), pp. 1–10, arXiv:151203980
21. O. Russakovsky, J. Deng, H. Su, J. Krause, S. Satheesh, S. Ma, Z. Huang, A. Karpathy, A. Khosla, M. Bernstein, A.C. Berg, L. Fei-Fei, ImageNet large scale visual recognition challenge. Int. J. Comput. Vis. **115**(3), 211–252 (2015)
22. M.S. Ryoo, L. Matthies, First-person activity recognition: feature, temporal structure, and prediction. Int. J. Comput. Vis. **119**(3), 307–328 (2016)
23. J. Shi, C. Tomasi, Good features to track, in *IEEE Conference on Computer Vision and Pattern Recognition* (IEEE, 1994), pp. 593–600
24. K. Simonyan, A. Zisserman, Two-stream convolutional networks for action recognition in videos, in *Advances in Neural Information Processing Systems* (2014), pp. 568–576
25. K. Simonyan, A. Zisserman, Very deep convolutional networks for large-scale image recognition, in *International Conference on Learning Representations* (2015), pp. 1–14
26. K. Soomro, A.R. Zamir, M. Shah, UCF101: a dataset of 101 human actions classes from videos in the wild (2012), pp. 1–7, arXiv:12120402
27. M.R. Souza, Digital video stabilization: algorithms and evaluation. Master's thesis, Institute of Computing, University of Campinas, Campinas, Brazil, 2018
28. L. Sun, K. Jia, K. Chen, D.Y. Yeung, B.E. Shi, S. Savarese, Lattice long short-term memory for human action recognition, in *IEEE International Conference on Computer Vision* (2017), pp. 2147–2156
29. S. Sun, Z. Kuang, L. Sheng, W. Ouyang, W. Zhang, Optical flow guided feature: a fast and robust motion representation for video action recognition, in *IEEE Conference on Computer Vision and Pattern Recognition* (2018), pp. 1390–1399

30. C. Szegedy, W. Liu, Y. Jia, P. Sermanet, S. Reed, D. Anguelov, D. Erhan, V. Vanhoucke, A. Rabinovich, Going deeper with convolutions, in *IEEE Conference on Computer Vision and Pattern Recognition* (2015), pp. 1–12
31. C. Szegedy, V. Vanhoucke, S. Ioffe, J. Shlens, Z. Wojna, Rethinking the inception architecture for computer vision, in *IEEE Conference on Computer Vision and Pattern Recognition* (2016), pp. 2818–2826
32. B.S. Torres, H. Pedrini, Detection of complex video events through visual rhythm. Vis. Comput. 1–21 (2016)
33. D. Tran, H. Wang, L. Torresani, J. Ray, Y. LeCun, M. Paluri, A closer look at spatiotemporal convolutions for action recognition, in *IEEE Conference on Computer Vision and Pattern Recognition* (2018), pp. 6450–6459
34. Z. Tu, W. Xie, J. Dauwels, B. Li, J. Yuan, Semantic cues enhanced multi-modality multi-stream CNN for action recognition. IEEE Trans. Circuits Syst. Video Technol. **29**(5), 1423–1437 (2018)
35. G. Varol, I. Laptev, C. Schmid, Long-term temporal convolutions for action recognition. IEEE Trans. Pattern Anal. Mach. Intell. **40**(6), 1510–1517 (2018)
36. L. Wang, Y. Qiao, X. Tang, Action recognition with trajectory-pooled deep-convolutional descriptors, in *IEEE Conference on Computer Vision and Pattern Recognition* (2015), pp. 4305–4314
37. L. Wang, Y. Xiong, Z. Wang, Y. Qiao, Towards good practices for very deep two-stream convnets (2015), pp. 1–5, arXiv:150702159
38. L. Wang, Y. Xiong, Z. Wang, Y. Qiao, D. Lin, X. Tang, L. Van Gool, Temporal segment networks: towards good practices for deep action recognition, in *European Conference on Computer Vision* (Springer, 2016), pp. 20–36
39. Y. Wang, M. Long, J. Wang, P.S. Yu, Spatiotemporal pyramid network for video action recognition, in *IEEE Conference on Computer Vision and Pattern Recognition* (IEEE, 2017), pp. 2097–2106
40. H. Wang, Y. Yang, E. Yang, C. Deng, Exploring hybrid spatio-temporal convolutional networks for human action recognition. Multimed. Tools Appl. **76**(13), 15065–15081 (2017)
41. J. Wang, A. Cherian, F. Porikli, S. Gould, Video representation learning using discriminative pooling, in *IEEE Conference on Computer Vision and Pattern Recognition* (2018), pp. 1149–1158
42. M.A. Wani, F.A. Bhat, S. Afzal, A.I. Khan, Advances in Deep Learning, vol. 57 (Springer, Berlin, 2020)
43. H. Yang, C. Yuan, B. Li, Y. Du, J. Xing, W. Hu, S.J. Maybank, Asymmetric 3D convolutional neural networks for action recognition. Pattern Recognit. **85**, 1–12 (2019)
44. W. Zhu, J. Hu, G. Sun, X. Cao, Y. Qiao, A key volume mining deep framework for action recognition, in *IEEE Conference on Computer Vision and Pattern Recognition* (IEEE, 2016), pp. 1991–1999
45. J. Zhu, Z. Zhu, W. Zou, End-to-end video-level representation learning for action recognition, in *24th International Conference on Pattern Recognition* (IEEE, 2018), pp. 645–650
46. Y. Zhu, PyTorch implementation of popular two-stream frameworks for video action recognition (2019), https://github.com/bryanyzhu/two-stream-pytorch

Deep Active Learning for Image Regression

Hiranmayi Ranganathan, Hemanth Venkateswara, Shayok Chakraborty and Sethuraman Panchanathan

Abstract Image regression is an important problem in computer vision and is useful in a variety of applications. However, training a robust regression model necessitates large amounts of labeled training data, which is time-consuming and expensive to acquire. Active learning algorithms automatically identify the salient and exemplar instances from large amounts of unlabeled data and tremendously reduce human annotation effort in inducing a machine learning model. Further, deep learning models like Convolutional Neural Networks (CNNs) have gained popularity to automatically learn representative features from a given dataset and have depicted promising performance in a variety of classification and regression applications. In this chapter, we exploit the feature learning capabilities of deep neural networks and propose a novel framework to address the problem of active learning for regression. We formulate a loss function (based on the expected model output change) relevant to the research task and exploit the gradient descent algorithm to optimize the loss and train the deep CNN. To the best of our knowledge, this is the first research effort to learn a discriminative set of features using deep neural networks to actively select informative samples in the regression setting. Our extensive empirical studies on five benchmark regression datasets (from three different application domains: rotation angle estimation of handwritten digits, age, and head pose estimation) demonstrate the merit of our framework in tremendously reducing human annotation effort to induce a robust regression model.

H. Ranganathan
Lawrence Livermore National Laboratory, 7000, East Avenue, Livermore, CA 94550, USA
e-mail: hrangana@asu.edu

H. Venkateswara · S. Panchanathan
Arizona State University, 699 S Mill Ave, Tempe, AZ 85281, USA
e-mail: hemanthv@asu.edu

S. Panchanathan
e-mail: panch@asu.edu

S. Chakraborty (✉)
Florida State University, 1017 Academic Way, Tallahassee, FL 32304, USA
e-mail: shayok@cs.fsu.edu

M. A. Wani et al. (eds.), *Deep Learning Applications*,
Advances in Intelligent Systems and Computing 1098,
https://doi.org/10.1007/978-981-15-1816-4_7

1 Introduction

Image regression is a well-researched problem in computer vision and is useful in a variety of applications, including age estimation [37], head pose estimation [35] and facial landmark detection [44] among others. A fundamental challenge in training reliable models for image regression is the requirement of large amounts of labeled training data. However, while gathering large quantities of unlabeled data is cheap and easy, annotating the data (with class labels) is an expensive process in terms of time, labor, and human expertise. Thus, developing algorithms to minimize human effort in training models for image regression is a fundamental research challenge. *Active Learning (AL)* algorithms automatically identify the salient and exemplar samples for manual annotation when exposed to large amounts of unlabeled data. This tremendously reduces the human annotation effort, as only a few samples, which are identified by the algorithm, need to be labeled manually. Further, since the model gets trained on the exemplar instances, its generalization capability is much better than that of a passive learner, which selects training examples at random. In recent years, active learning has been used in many machine learning applications [7, 13] with promising results. While active learning has been extensively studied for classification [27, 38, 46], active learning for regression is much less explored.

Deep Learning (DL) algorithms have recently emerged as a dominant machine learning tool to learn representative features for classification and regression tasks [26] and have replaced the need for handcrafted features. Architectures such as Convolutional Neural Networks (CNNs), Recurrent Neural Networks (RNNs), etc., have created a paradigm shift in multimedia computing applications. Deep learning has been widely explored in computer vision and has achieved tremendous performance improvements in several vision tasks, including image recognition [24], object detection [16], multimodal emotion recognition [31, 32] and image segmentation [28] among others. Besides classification, CNNs have also been effectively exploited for regression tasks such as pose estimation [47], object detection [45], facial landmark detection [44], and depth prediction [12].

In this chapter, we exploit the virtue of deep networks to learn rich sets of features and develop a novel active learning framework for the regression setting. We use the Expected Model Output Change (EMOC) as the active selection criterion and integrate it within the objective function used to train the deep learning model. The resulting model optimizes this novel objective and learns from salient examples that cause maximum change to the current model. Research in deep active learning is still in a nascent stage [33, 34, 41, 48, 58]. To the best of our knowledge, this is the first research effort to develop an active learning algorithm for regression using deep convolutional neural networks. Although validated on image regression in this research, the proposed framework is generic and can be used in any regression application where the most informative examples need to be selected for manual annotation, from large amounts of unlabeled data.

The rest of the chapter is organized as follows: we present a survey of related techniques in Sect. 2; Sect. 3 details the proposed deep active learning framework for regression; the experiments and results are presented in Sect. 4 and we conclude with discussions in Sect. 5.

2 Related Work

In this section, we present a brief survey of existing work in deep learning and active learning for regression.

2.1 Deep Learning for Regression

Deep learning has been successfully applied in a variety of regression applications, including object (crowd) counting, age estimation from face images, human pose estimation, and depth estimation among others. Shi et al. proposed a deep learning algorithm for crowd counting, which learns generalizable features using a negative correlation [40]. Zhao et al. addressed the challenging problem of estimating the number of people across a line of interest in surveillance videos, using CNNs [57]. Zhang et al. proposed a cross scene crowd counting technique using deep CNNs for counting people in target surveillance crowd scenes, which are unseen in the training set [56]. Liu et al. used CNNs together with Long Short Term Memory (LSTM) networks for crowd counting in a closed environment with WIFI signals [29]. Age estimation from facial images is another computer vision-based regression application where deep learning has shown a remarkable success. Rothe et al. proposed a deep learning solution to age estimation from a single face image without the use of facial landmarks [36, 37]. The authors also introduced the IMDB-Wiki dataset, which is the largest public dataset of face images with age and gender labels. Wang et al. used feature maps obtained in different layers of a deep learning model for age estimation, instead of using the features obtained only in the last layer [49]; together with a manifold learning algorithm, their framework depicted significant performance improvements over the baselines. Zaghbani et al. exploited autoencoders to learn features in a supervised manner to estimate age from face images [54]. CNNs have been successfully applied for human pose estimation, where the regressed values correspond to the positions of the body joints on the image plane [30, 47]. Sun et al. used CNNs effectively to predict the facial fiducial points in facial landmark detection [44]. Gkioxari et al. used R-CNNs with a loss function composed of a body pose estimation term and an action detection term [17]. Szegedy and Jaderberg et al. used deep networks for object and text detection to predict a bounding box for localization [22, 45]. The above deep models use the conventional L_2 loss function for training. Zhang et al. introduced a CNN optimized for landmark detection and attribute classification [55]. They combined the standard L_2 loss function with the softmax classification function to increase robustness to outliers. Wang et al. combined bounding box localization with object segmentation using a similar approach [50]. Dosovitskiy and Eigen et al. used multiple L_2 loss functions for object generation and depth estimation [11, 12]. From the above survey, we see that deep models (specifically CNNs) trained using the L_2 loss function can be applied effectively for regression tasks. Therefore, we use CNNs as our preferred deep model in this work.

2.2 Active Learning for Regression

Research in active learning for regression is much less explored compared to active learning methods developed for classification. Willett et al. provided a theoretical analysis of active learning in the context of regression [52]. Population based active learning methods were proposed by Sugiyama using the weighted least-squares learning, where they predicted the conditional expectation of the generalization error given the input training samples [42]. A theoretically optimal active learning algorithm was proposed by Sugiyama and Nakajima [43]. This directly minimized the generalization error by employing an additive regression model. Freund et al. applied a variance-based Query-by-Committee (QBC) framework to regression [14]. Cohn et al. minimized the output variance to reduce the generalization error [10]. Yu and Kim provided passive sampling heuristics based on the geometric characteristics of the data [53]. Burbidge et al. investigated a committee-based approach for active learning of real-valued functions. They used a variance-only strategy for the selection of informative training data [5]. Freytag et al. proposed an approach to measure the expected change of model outputs [15]. For each example in the unlabeled set, the expected change of model predictions was calculated and marginalized over the unknown label. The resulting score for each unlabeled example was used for active learning with a broad range of models and learning algorithms. Most regression-based active learning techniques are developed only for the sequential query mode (querying only a single unlabeled sample in each iteration). Batch Mode Active Learning (BMAL) techniques, which query batches of unlabeled samples simultaneously, are very useful in practice as they can exploit the presence of multiple labeling oracles. While BMAL has been extensively studied in the context of classification [1, 3, 4, 8, 9, 18–21], it has been much less explored in the regression setting. Cai et al. extended sequential mode of active learning to BMAL by simulating the sequential mode active learning behavior to simultaneously choose a set of examples without retraining [6]. They introduced a novel active learning framework for regression called Expected Model Change Maximization (EMCM), which queried the examples maximizing the model change once added to the training data. Along similar lines, Käding et al. proposed an active learning algorithm based on the expected model output change, which queried samples that could potentially impart the maximal change in the model output, if added to the training set [23].

Even though deep learning and active learning for regression have been studied individually, research on developing an end to end deep active learning framework for regression is unexplored. In this chapter, we exploit the feature learning capabilities of deep networks and propose a novel framework to address the problem of active learning for regression. We use Expected Model Output Change (EMOC) as the active selection criterion and integrate it within the objective function used to train the deep model. Since the active learning criterion is embedded within the loss function, the network gets specifically trained for the task of active learning and can potentially depict better performance than a network trained merely using a conventional L_2 loss. Furthermore, our technique considers changes in all the parameterized layers of the deep network and implicitly combines them into a single criterion for active sample selection. This is in contrast to existing methods that only make use of the current output of a deep neural network for querying unlabeled samples. We now describe our framework.

3 Proposed Framework

In our active learning setting, the deep CNN is exposed to a small amount of labeled data and a large amount of unlabeled data. In each iteration of our algorithm, a batch of unlabeled samples is selected and given to the human oracle for annotation. These labeled samples are removed from the unlabeled set and appended to the labeled set. The deep model is retrained on the updated sets and evaluated on a heldout test set. The process is repeated until some stopping criterion is satisfied (defined as a pre-determined number of iterations in this work). The challenge in active learning is to identify the most informative set of unlabeled samples to be labeled by the oracle, so as to obtain the maximal performance on the test set with minimal human effort.

The core idea of this research is to leverage the feature learning capabilities of deep neural network models to identify the most informative unlabeled samples for active learning. We attempt to integrate an active sample selection criterion in the objective function and train the network to minimize this objective. The features learned by the network are then specially tailored to the active learning task. This enables the model to better identify the samples that can augment maximal information to the model. We use the EMOC criterion to quantify the utility of an unlabeled sample in our active learning framework. We achieve this by adding an EMOC based loss term to the conventional regression objective and train the network to optimize this joint objective function.

Formally, let $g(x; \phi)$ be the output of a neural network where parameters ϕ are the parameters of the network and x is the input image. In this work, we focus on layered deep models, $g(x_i; \phi) = g_l(\cdots(g_2(g_1(x_i; \phi_1); \phi_2)\cdots); \phi_l)$. Here, $\phi = (\phi_1, \ldots, \phi_l)$ denotes the parameters of the deep model and l is the total number of layers in the deep model. Let the set of labeled samples be represented as $X^L = \{x_1, x_2, \ldots, x_{n_l}\}$. The corresponding labels for X^L are denoted by $Y^L = \{y_1, y_2, \ldots, y_{n_l}\}$ representing continuous real values; $y_i \in \mathbb{R}$. Let the set of unlabeled samples be $X^U = \{x_{n_l+1}, x_{n_l+2}, \ldots, x_{n_l+n_u}\}$. Let $X = X^L \cup X^U$ denote the union of the disjoint subsets X^L and X^U and $n = n_l + n_u$. The goal of active learning is to select a batch B containing k unlabeled samples for manual annotation such that the modified learner (trained on labeled set $X^L \cup B$ and unlabeled set $X^U \backslash B$) has maximum generalization capability. We now formulate a novel loss function to train the deep CNN for active learning. Our loss function consists of two terms—one term to quantify the loss on the labeled data and a second term to quantify the loss on the unlabeled data. These are detailed below.

3.1 Loss on Labeled Data

The purpose of this term is to ensure that the trained CNN has minimal prediction error on the labeled data. We use the conventional L_2 loss to quantify the error on the labeled data. Consider a subset of labeled samples $X^l = \{x_1, x_2, \ldots, x_{n_l'}\}$ and their corresponding labels $Y^l = \{y_1, y_2, \ldots, y_{n_l'}\}$. Let $\hat{Y}^l = [\hat{y}_1, \hat{y}_2, \ldots, \hat{y}_{n_l'}] = [g(x_1; \phi), g(x_2; \phi), \ldots, g(x_{n_l'}; \phi)]$, be the predictions of the deep neural network on the subset of labeled data. The prediction loss is given by

$$\mathcal{L}(\phi; X^l, Y^l) = \frac{1}{n_l'} \sum_{i=1}^{n_l'} \left(y_i - \hat{y}_i\right)^2 \tag{1}$$

Minimizing this loss ensures that the trained model makes predictions that are consistent with the labeled data.

3.2 Principle of Expected Model Output Change (EMOC)

The EMOC criterion provides a principled way to quantify the importance of a sample by measuring the difference of model outputs when trained with and without a particular data sample:

$$\Delta g(x') = \mathbb{E}_{y'|x'} \mathbb{E}_x \left|\left|g(x; \phi') - g(x; \phi)\right|\right|_1 \tag{2}$$

In Eq. (2), $||g(x; \phi') - g(x; \phi)||_1$ computes the L_1 norm of the difference between the outputs of the models. Here, ϕ' denotes the parameters of the model obtained by additionally training with unlabeled example x'. In order to estimate ϕ' we need to know the label of x'. We assume y' is the label for x'. In general, the first expectation operation is used to marginalize over y' in the above equation to get the expected model change. The expectation $\mathbb{E}_x$ is estimated by computing the empirical mean across the dataset and the expectation $\mathbb{E}_{y'|x'}$ is based on the output of the updated model $g(.; \phi')$ for all possible values of y' given x'.

A direct implementation of the EMOC principle would require training a model from scratch for each example x' in the dataset, making it very computation intensive. Therefore, development of efficient techniques that approximate the change in model output $\Delta g(.)$, is required. Freytag et al. derived a closed form expression for $\Delta g(x')$ focusing on Gaussian process regression [15]. Käding et al. used the stochastic gradient approximation with a single sample to estimate model parameter updates [23]. This approximation is given in Eq. (3), where the gradient of the objective with respect to a candidate example (x', y') is used to estimate the model changes:

$$(\phi' - \phi) \approx \eta \nabla_{\phi} \mathcal{L}(\phi; (x', y')), \tag{3}$$

where, $\eta > 0$ is some constant. The difference $||g(x; \phi') - g(x; \phi)||_1$ can be approximated using the first-order Taylor series approximation as

$$\big|\big|g(x;\phi') - g(x;\phi)\big|\big|_1 \approx \big|\big|\nabla_{\phi} g(x;\phi)^{\top}(\phi' - \phi)\big|\big|_1 \tag{4}$$

We substitute Eq. (3) in Eq. (4) to get:

$$||g(x;\phi') - g(x;\phi)||_1 \approx \eta \big|\big|\nabla_{\phi} g(x;\phi)^{\top} \nabla_{\phi} \mathcal{L}(\phi; (x', y'))\big|\big|_1 \tag{5}$$

Since marginalizing over all possible values for y' is impractical, an approximation was proposed which considers only the most likely label $\bar{y}'$, (as the label for unlabeled sample x'), inferred by the model g [23]. It is, therefore, assumed that all examples in a given unlabeled set X' have label $\bar{y}'$. With this simplifying approximation, the EMOC score for an unlabeled set X' is given by

$$\Delta g(X') = \sum_{x' \in X'} \mathbb{E}_x \big|\big|\nabla_{\phi} g(x;\phi)^{\top} \nabla_{\phi} \mathcal{L}(\phi; (x', \bar{y}'))\big|\big|_1 \tag{6}$$

3.3 Loss on Unlabeled Data

The purpose of this term is to ensure that the trained CNN has maximal prediction confidence on the unlabeled samples. In a classification setting, the prediction confidence of a trained network on an unlabeled sample is usually quantified using the Shannon's entropy. The entropy of each unlabeled sample is computed from the probability distribution of the network outputs on that sample and the total entropy on the unlabeled set is included as a term in the overall loss function [33, 34]. Minimizing this term ensures that the network is trained such that it furnishes high confidence (low entropy) predictions on the unlabeled set. However, computing the prediction uncertainty in a regression application is a challenge, as entropy does not have an exact analog in the regression setting. We, therefore, leverage the principle of EMOC to compute the prediction uncertainty of the deep model on a given unlabeled sample. We impose a condition in the loss function which ensures that all the unlabeled samples have low EMOC scores with respect to the trained model. Thus, no unlabeled sample can drastically affect the model parameters, i.e., the model depicts high prediction confidence on the unlabeled set. Further, we see the following benefits to incorporate the unlabeled data when training the network: (i) the CNN is trained to extract features from both the labeled and unlabeled samples, making it more robust compared to a network that is only trained on labeled samples; (ii) the EMOC loss from unlabeled data acts as a regularizer, preventing over-fitting and improving the generalization capabilities of the network; and (iii) since the network minimizes the EMOC loss, this helps in selecting the most relevant samples to form the batch B.

Let $X^u = \{x'_1, x'_2, \ldots, x'_{n'_u}\}$, denote a subset of unlabeled samples. We do not have the labels for X^u. Estimating the model change in Eq. (2), by marginalizing over all possible labels for X^u is computation-intensive and impractical. We, therefore, approximate the labels of X^u to be the mean of the labels inferred by model g, along the lines of [23]. We forward propagate $X^u = \{x'_1, x'_2, \ldots, x'_{n'_u}\}$ through the network $g(.; \phi)$ and obtain the predictions $\hat{Y}^u = \{\hat{y}'_1, \hat{y}'_2, \ldots, \hat{y}'_{n'_u}\}$. Let $\bar{y}' = 1/n'_u \sum_{i=1}^{n'_u} \hat{y}'_i$ be the mean of $\hat{Y}^u$. The unlabeled samples and their approximated labels are: $\{(x'_1, \bar{y}'), (x'_2, \bar{y}'), \ldots, (x'_{n'_u}, \bar{y}')\}$. We formulate Eq. (6) as a loss on unlabeled data. Therefore, in our framework, the loss over a subset of unlabeled data is given by

$$\mathcal{U}(\phi; X^u) = \sum_{x' \in X^u} \mathbb{E}_x \left|\left| \nabla_\phi g(x; \phi)^\top \nabla_\phi \mathcal{L}(\phi; (x', \bar{y}')) \right|\right|_1 \tag{7}$$

Here, $\mathcal{L}(\phi; (x', \bar{y}')) = (\bar{y}' - \hat{y}')^2$, where x' is an unlabeled sample, $\bar{y}'$ is its approximate label, $\hat{y}'$ is the output of the network and $\mathbb{E}_x$ is the empirical mean over the dataset. Minimizing this loss ensures that the features are learned in such a way that all the unlabeled samples have low EMOC scores with respect to the trained model.

3.4 Novel Joint Objective Function

The deep model is trained using both labeled and unlabeled data with the objective of minimizing the L_2 loss on labeled data and the EMOC loss on the unlabeled data. The joint loss ensures that the network can accurately predict the labels of the labeled training data while the unlabeled samples have minimal effect on the trained model parameters; i.e., the model depicts good performance on the unlabeled data. Over successive iterations, the positive effects of this joint training with labeled and unlabeled data get enhanced. Our novel joint objective function over a batch of labeled and unlabeled data is given by

$$\mathcal{J}(\phi, X^l, Y^l, X^u) = \mathcal{L}(\phi; X^l, Y^l) + \lambda \mathcal{U}(\phi; X^u) \tag{8}$$

Here, $\lambda \geq 0$ controls the relative importance of the two terms. The objective function in Eq. (8) is minimized over multiple batches of labeled and unlabeled data using mini-batch gradient descent. In order to train our network, we compute $\nabla_\phi \mathcal{J}$ and use back-propagation to update the network parameters (the next section gives the expression for the gradient $\nabla_\phi \mathcal{J}$). Once the network is trained, the unlabeled examples with the largest EMOC scores are selected to form the batch B. These samples are annotated by a human expert and the resulting labeled batch is appended to the labeled set X^L. Since the network is trained to minimize the EMOC score of the unlabeled samples, the unlabeled samples furnishing the highest EMOC scores after model training are the most informative data points with respect to the current model. They are hence queried for labels.

3.5 Gradient of the Objective Function

We provide a high-level overview of the gradient computation of the objective function in this section.

The gradient of the joint objective function for deep active regression is given by

$$\nabla_\phi \mathcal{J}(\phi, X^l, Y^l, X^u) = \nabla_\phi \mathcal{L}(\phi; X^l) + \lambda \nabla_\phi \mathcal{U}(\phi; X^u), \tag{9}$$

where the gradients for individual layers are

$$\frac{\partial \mathcal{L}(\phi; X^l)}{\partial \phi_j} = \frac{-2}{n_l'} \sum_{i=1}^{n_l'} \left[(y_i - \hat{y}_i) \frac{\partial g(x_i; \phi)}{\partial \phi_j} \right], \ \forall j \in \{1, 2, \ldots, l\}, \tag{10}$$

and

$$\begin{aligned} \frac{\partial \mathcal{U}(\phi; X^u)}{\partial \phi_j} &= \sum_{x' \in X^u} \mathbb{E}_x \frac{\partial}{\partial \phi_j} \left|\left| \nabla_\phi g(x; \phi)^\top \nabla_\phi \mathcal{L}(\phi; (x', \bar{y}')) \right|\right|_1 \\ &= \sum_{x' \in X^u} \mathbb{E}_x \, (Q_j), \quad \forall j \in \{1, 2, \ldots, l\} \end{aligned} \tag{11}$$

In Eq. (11), Q_j stands for,

$$Q_j = \begin{cases} +(Q_{j_1} + Q_{j_2}) & \text{if} \quad \nabla_\phi g(x; \phi)^\top \nabla_\phi \mathcal{L}(\phi; (x', \bar{y}')) \geq 0 \\ -(Q_{j_1} + Q_{j_2}) & \text{if} \quad \nabla_\phi g(x; \phi)^\top \nabla_\phi \mathcal{L}(\phi; (x', \bar{y}')) < 0, \end{cases} \tag{12}$$

with

$$Q_{j_1} = \frac{\partial}{\partial \phi_j} \left(\nabla_\phi g(x; \phi)^\top \right) \nabla_\phi \mathcal{L}(\phi; (x', \bar{y}')), \tag{13}$$

and

$$Q_{j_2} = \nabla_\phi g(x; \phi)^\top \frac{\partial}{\partial \phi_j} \left(\nabla_\phi \mathcal{L}(\phi; (x', \bar{y}') \right) \tag{14}$$

We compute $Q_{j_1} + Q_{j_2}$ for a fixed $x' \in X^u$ and for all $x \in X^l$. Then we compute the expected value $\mathbb{E}_x(Q_j)$. We do this for every $x' \in X^u$ and then compute $\sum_{x' \in X^u} \mathbb{E}_x(Q_j),\ \forall j \in \{1, 2, \ldots, l\}$ to get $\nabla_\phi \mathcal{U}(\phi; X^u)$.

Algorithm 1 The proposed Deep Active Algorithm for Regression

Require: The labeled set X^L, labels Y^L, unlabeled set X^U, weight parameter λ, batch size k, maximum number of iterations T

1: **for** $t = 1, 2, \ldots T$ **do**
2: Compute the derivative of the joint objective function as in Eq. (9)
3: Train the deep model to optimize the joint loss
4: Compute the EMOC score of each unlabeled sample, using Eq. (6)
5: Select a batch B containing k unlabeled samples from X_u furnishing the highest EMOC scores
6: Update $X^L \rightarrow X^L \cup B$; $X^U \rightarrow X^U \backslash B$
7: **end for**

Figure 1 shows a graphical illustration of the proposed framework. We present the network with a mini-batch of n' data points consisting of n'_l labeled points and n'_u unlabeled points; $n' = n'_l + n'_u$, $(n'_l \leq n_l, n'_u \leq n_u)$. The L_2 loss is computed over the labeled data in the mini-batch and the EMOC loss is computed over the unlabeled data in the mini-batch. The negative gradient of the joint objective function with respect to the mini-batch is back-propagated to train the CNN. The weight parameter λ was selected to be 1 giving equal weightage to both the terms. When the network has seen all the data points in the training set (both labeled and unlabeled), we consider it as one epoch. We repeat the training procedure over multiple epochs until convergence and consider this one training iteration t of the active learning algorithm. At the end of every iteration t, we sample the most informative batch of unlabeled data samples (samples furnishing highest EMOC score from Eq. (6)) to form B. We

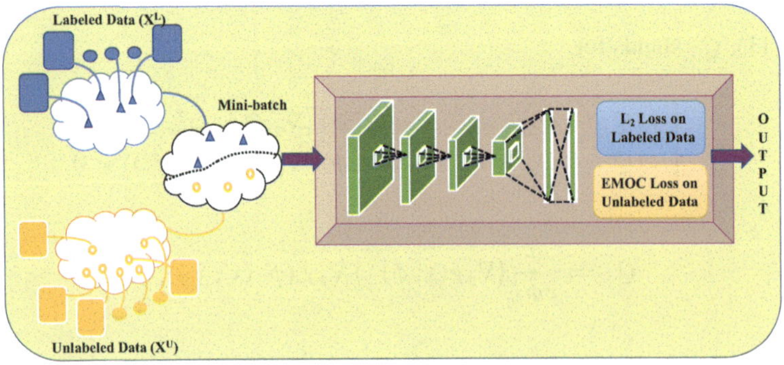

Fig. 1 Illustration of the proposed deep active learning framework. The deep model is trained with a combination of labeled and unlabeled mini-batches to minimize the L_2 loss over labeled data and EMOC loss over unlabeled data. A batch B of unlabeled points furnishing maximum EMOC scores is annotated and added to the labeled set. Best viewed in color

obtain the labels for B using an oracle and update the labeled and unlabeled datasets as discussed earlier. We iterate until we run out of unlabeled data points to be labeled or run out of budget to get them labeled. For implementation purposes, we fix the maximum number of iterations as T. The pseudocode of the proposed algorithm is given in Algorithm 1.

We note that an outlier in the dataset also furnishes a high EMOC score. Therefore, a significant change to the current model output does not always lead to better generalization performance. However, when the model has been altered by an outlier, the joint training objective along with the EMOC sampling criterion selects an informative set of examples in the next iteration that instantly alleviates the adverse effect of the outlier. In general, the number of outliers is low compared to the number of samples in the training set. Hence, it is reasonable to assume that the proposed framework will result in good generalization performance when salient examples are added to the labeled set over time.

4 Experiments and Results

4.1 Implementation Details

The proposed algorithm can be implemented in conjunction with any deep learning architecture. In this research, we studied the performance of our framework using Convolutional Neural Networks (CNNs) due to their popularity in computer vision applications [51]. Figure 2 illustrates the network architecture of the CNN used for deep active regression. As seen in Fig. 2, the size of the input image in the input layer of the CNN is 128×128 pixels. The convolution layer of our CNN model performs convolution operations with a kernel size of 3×3 pixels to acquire feature maps of the input information. The dimension of the first convolution layer is $128 \times 128 \times 32$ which denotes a feature size of 128×128 pixels and 32 different convolution kernels. All convolution layers are connected to RELU activation functions and max-pooling layers. The dimensions of the second, third, and fourth convolution layers are $64 \times 64 \times 64$, $32 \times 32 \times 128$ and $16 \times 16 \times 256$, respectively. The dimensions of each fully connected layer are 2048. The activation function of the output layer is a linear function so as to obtain a continuous value output. The network is trained by minimizing the joint loss function given in Eq. (8) using mini-batch gradient descent with an initial learning rate of 0.01. The implementations were performed in Matlab R2017b on a machine running a NVIDIA 1080Ti GPU with 11 GB memory.

4.2 Datasets and Experimental Setup

We used five benchmark regression datasets to evaluate our deep active framework: (i) *Synthetic handwritten digit dataset* [59]; (ii) *Rotated MNIST Digits* [25]; (iii) *WIKI Age Estimation Dataset* [36]; (iv) *BIWI Kinect Dataset* [2] and (v) the *QMUL*

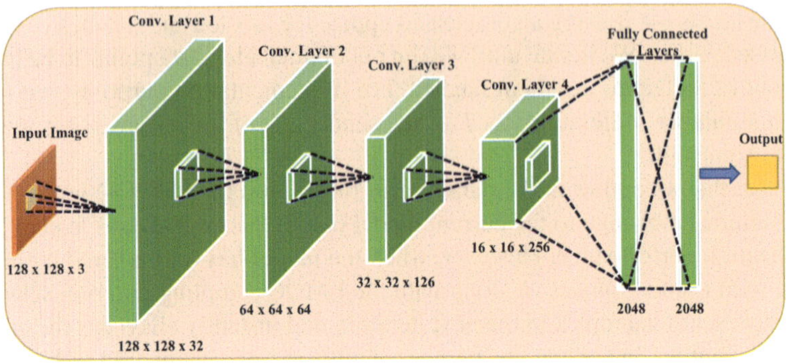

Fig. 2 CNN architecture for deep active regression

Table 1 Dataset details

Dataset name	Labeled (X^L)	Unlabeled (X^U)	Test set ($\mathcal{T}$)	Batch size (k)
Synthetic handwritten digits	500	4500	1000	200
WiKI age estimation	20000	30000	10000	400
MNIST rotation	15000	25000	5000	400
BIWI kinect	4000	6000	4000	400
QMUL multiview	800	4200	1000	200

Multiview Face Dataset [39]. These datasets represent different application domains (head pose, age, and handwritten digit recognition) and have been widely used for validating the performance of regression models. Our objective was to test the performance of the proposed active sampling framework for deep learning and not to outperform the best accuracy results on these datasets; so, we did not follow the precise train/test splits given for these datasets. We split each dataset into three disjoint parts to construct the initial labeled set X^L, unlabeled set X^U, and the test set $\mathcal{T}$. Each algorithm (baseline and proposed) selected k instances from the unlabeled pool to be labeled in each iteration. After each iteration, the selected samples were removed from the unlabeled set, appended to the training set and the performance was evaluated on the test set. The goal was to study the improvement in performance on the test set with increasing sizes of the training set. The experiments were run for 15 iterations. The dataset details are summarized in Table 1.

4.3 Comparison Baselines and Evaluation Metrics

To study the performance of our proposed framework, we compared our method against four state-of-the-art regression-based active learning algorithms: (i) **Käding** [23]: In this method a CNN, described in Sect. 4.1, is trained using the L_2 loss function given in Eq. (1). Unlabeled samples with the largest EMOC are selected to form a batch. Note that, this is a two-step process, where a CNN is first trained using a conventional loss function and the EMOC criterion is then applied for active sampling. In contrast, our framework integrates the EMOC criterion in the loss function to train the network; *(ii)* **Greedy**: This model selects unlabeled examples having the largest minimum distance from labeled data [53]; *(iii)* **Query-by-Committee (QBC)**: This model selects data points that have the largest variance among the committee's predictions [5]; and *(iv)* **Random Sampling**: This method selects a batch of samples at random from the unlabeled pool. The *QBC* and *Greedy* active learning strategies were not proposed in the context of deep learning. However, for a fair comparison, we trained the CNN described in Sect. 4.1 using the standard L_2 loss function and then applied the active selection criterion.

For evaluation, we used two popular error-based metrics, Mean Squared Error (*MSE*), and Mean Absolute Error (*MAE*), to study the performance of each method on the test set

$$MSE = \frac{1}{|\mathcal{T}|} \sum_{i=1}^{|\mathcal{T}|} (y_i - g(x_i))^2 \tag{15}$$

$$MAE = \frac{1}{|\mathcal{T}|} \sum_{i=1}^{|\mathcal{T}|} |y_i - g(x_i)| \tag{16}$$

Here, $|\mathcal{T}|$ denotes the size of the test set; y_i and $g(x_i)$ are the ground truth and the predictions of test sample x_i, respectively.

4.4 Active Learning Performance

The performance of the five active learning algorithms on the benchmark datasets are presented in Figs. 3, 4, 5, 6, and 7. The x-axis represents the iteration number and the y-axis denotes the *MSE* values. In general, we see that the *MSE* values decrease when the number of training points increases for all five algorithms. This is in accordance with the insight that the performance of the model increases with an increase in labeled data. Our proposed deep active framework consistently depicts the best performance across all datasets; at any given iteration number, it has the least error among all the methods. This shows that the proposed framework can appropriately identify the most informative unlabeled samples for manual annotation and can attain a given performance level with the least human effort. The performance of the *Käding* [23] baseline is better than the other three baselines but is not as good as our method. This corroborates the fact that training a deep network to minimize a joint loss function containing the EMOC criterion depicts better performance than the two-step process of training a network to minimize the L_2 loss and then selecting samples based on EMOC. The *QBC* and *Greedy* algorithms outperform *Random Sampling*, but perform poorly compared to the *Käding* [23] baseline.

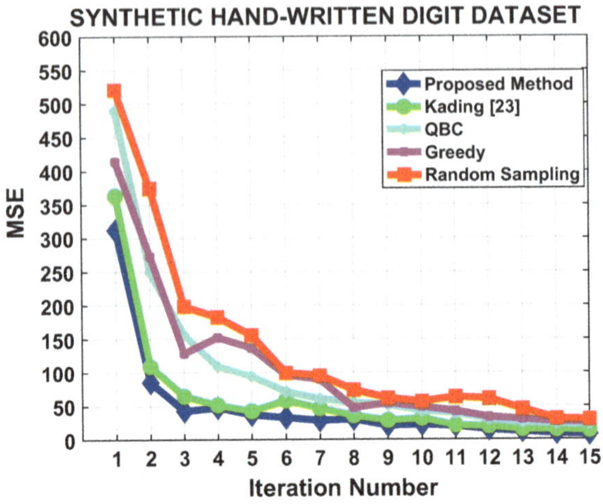

Fig. 3 MSE versus iteration number: synthetic handwritten digits. Best viewed in color

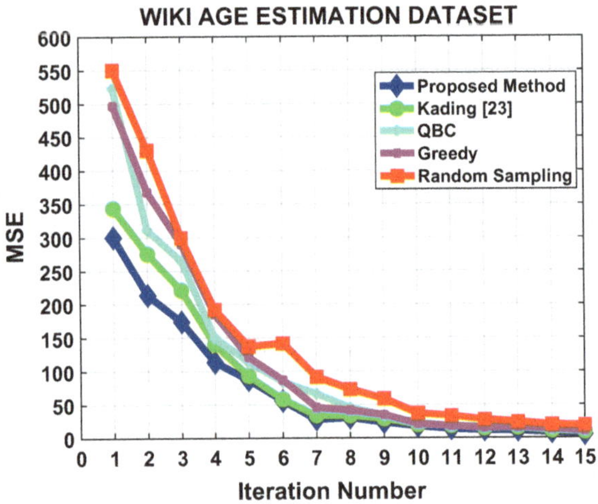

Fig. 4 MSE versus iteration number: WIKI age estimation. Best viewed in color

The *MAE* active learning curves for all the five datasets depict similar trends as the *MSE* curves. For the sake of brevity, we report the label complexity values in Table 2 (corresponding to $MAE = 9$). Each entry in the table denotes the number of unlabeled samples that had to be annotated to achieve an *MAE* value of 9. The results follow a similar pattern as in Figs. 3, 4, 5, 6, and 7; the proposed method requires the least amount of labeled samples (and consequently, the least human effort) to attain the given level of performance, for all the datasets. For the QMUL dataset, *Random Sampling* does not attain an *MAE* of 9 even after 15 iterations. The results unanimously lead to the conclusion that the proposed method consistently depicts the best performance over all the baseline algorithms, across all datasets.

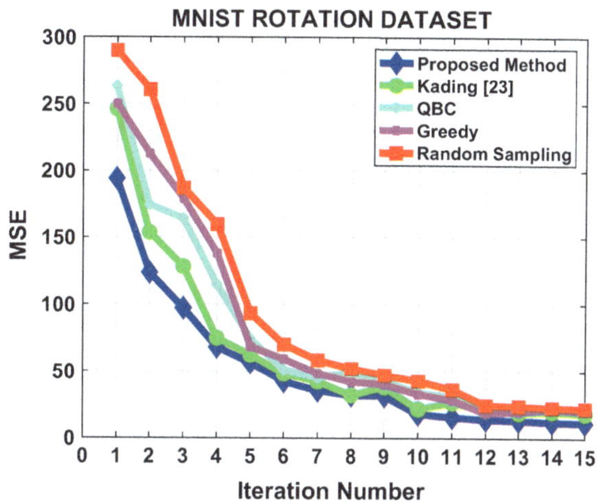

Fig. 5 MSE versus iteration number: MNIST rotation. Best viewed in color

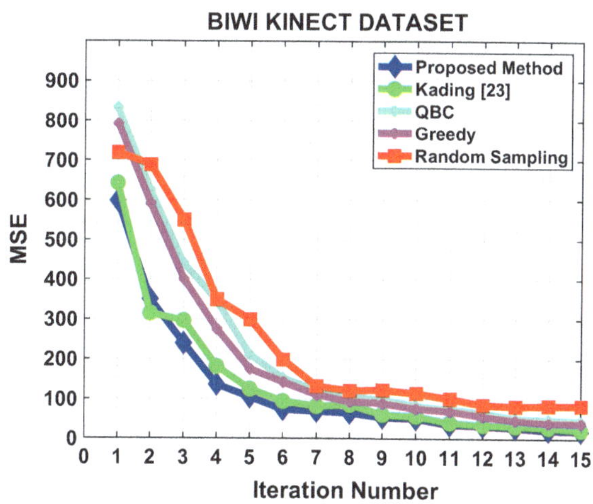

Fig. 6 MSE versus iteration number: BIWI kinect. Best viewed in color

4.5 Study of the Active Sampling Criterion

In order to further evaluate the active sampling criterion employed in our framework, we performed the following two experiments. We conducted these experiments on the MNIST Rotation dataset.

Experiment 1: We selected 200 samples of each digit (0–9) at random from the test set ($200 \times 10 = 2000$ samples). Figure 8a shows the performance of the proposed model after a randomly selected iteration (iteration number 9) per digit class on the selected 2000 samples. The x-axis corresponds to the digit class and y-axis shows the *MSE*. We perform similar experiments using the *Random Sampling* method. The performance of *Random Sampling* per digit class after iteration number 9 is shown

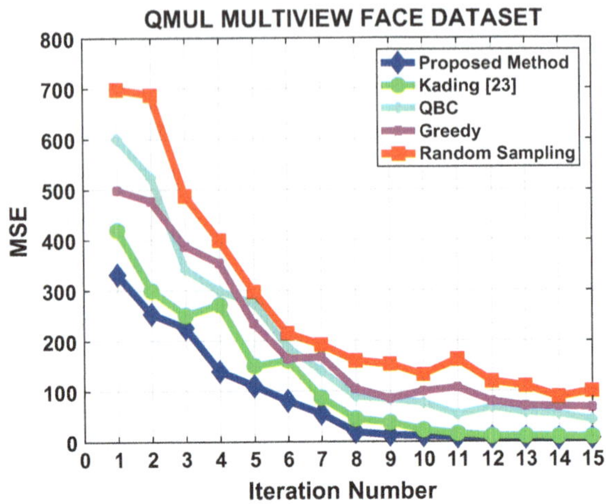

Fig. 7 MSE versus iteration number: QMUL multiface. Best viewed in color

Table 2 Label Complexity for $MAE = 9$. The proposed framework requires the least amount of labeled data to reach a given performance level ($MAE = 9$) compared to all the baseline methods. For the QMUL dataset, *Random Sampling* does not attain an *MAE* of 9 even after 15 iterations

Dataset name	Proposed method	Käding [23]	Greedy	QBC	Random sampling
Synthetic handwritten digits	400	600	1000	1000	1200
WiKI age estimation	1200	1600	2000	2400	2800
MNIST rotation	1200	1600	2400	2000	2800
BIWI kinect	2000	2400	4000	3200	5200
QMUL multiview	1000	1400	1600	2000	–

in Fig. 8b. We then plot the number of samples of each digit picked to form batch *B* after iteration number 9 using the proposed method and *Random Sampling*. The results are shown in Fig. 8c, d, respectively.

Observations: From Fig. 8a, we see that, the top four digits furnishing the maximum errors are digits 0, 4, 6, and 1 when using the proposed model. Similarly, from Fig. 8b, we observe that the top four digits furnishing the maximum error are 1, 5, 0, and 9 when using *Random Sampling*. From Fig. 8c, we observe that 65% of the 400 samples selected to form batch *B* by the proposed method, belonged to digits 0, 4, 6, and 1 (the digits furnishing the maximum error using the model after iteration number 9). This shows that our proposed model intelligently selects samples to augment the training set, which can maximally reduce the generalization error. On the other hand, when using *Random Sampling* (Fig. 8d), we notice that only 34.75% of the 400 samples selected to form batch *B* belonged to the four classes furnishing maximum error. This shows that there is no correlation between the number of samples selected for a digit to its corresponding error when using *Random Sampling*, accounting for its poor performance.

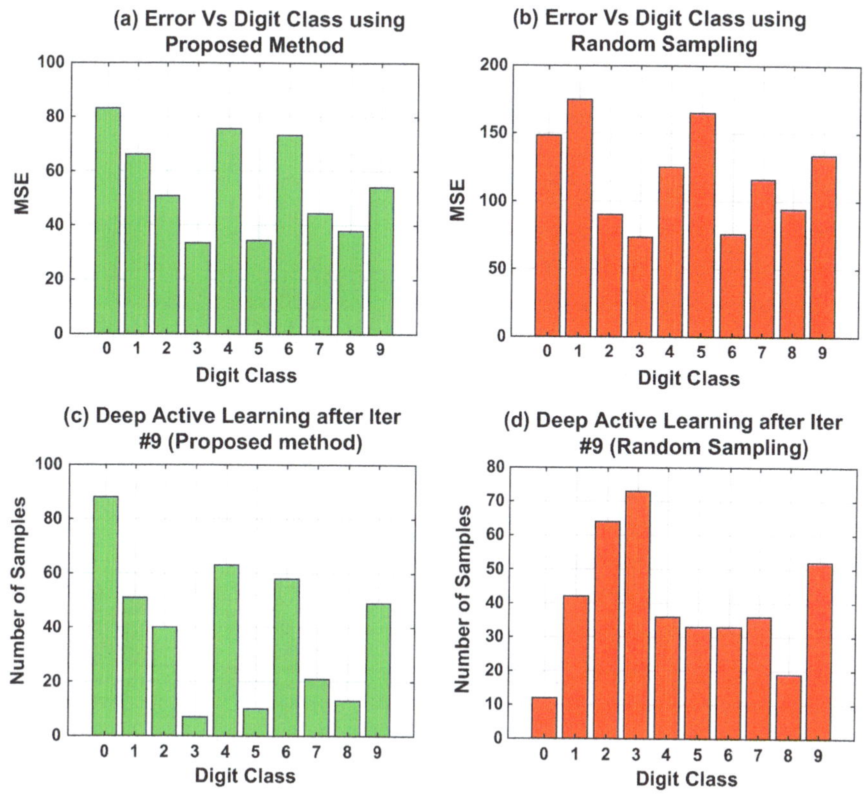

Fig. 8 Results after iteration number 9. **a** MSE versus digit class using proposed method, **b** MSE versus digit class using random sampling, **c** Number of samples of each digit (0–9) selected using proposed method and **d** Number of samples of each digit (0–9) selected using random sampling. Best viewed in color

Experiment 2: In this experiment, we further investigate the rotation angle of the four digits furnishing maximal errors from the previous experiment. The first row in Fig. 9 shows the performance of the proposed model per angular bin after iteration 9 (for the four digits 0, 4, 6, and 1 furnishing the maximal errors). We split the range of the predicted angles into 12 different bins (**Bin 1** : $(-60^\circ$ to $-50^\circ)$, **Bin 2** : $(-49^\circ$ to $-40^\circ)$, ..., **Bin 12** : $(+50^\circ$ to $+60^\circ)$). In the second row in Fig. 9, we plot the number of samples picked after iteration number 9 in each angular bin. The first and second rows in Fig. 10 show similar plots using the *Random Sampling* method (for the four digits 1, 5, 0 and 9 furnishing the maximal errors).

Observations: From Fig. 9, when using the proposed method, we see that there is a direct correlation between the angular bins showing high error and the number of samples chosen in those angular bins. We see no such relations when using *Random Sampling* in Fig. 10. This further corroborates the usefulness of the active sampling criterion used to train the deep CNN in our framework.

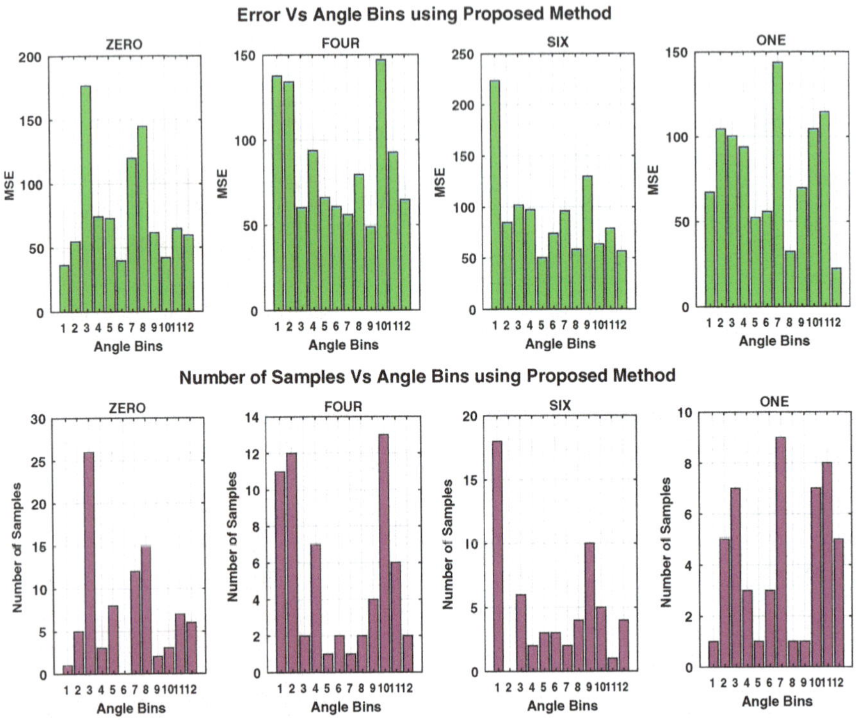

Fig. 9 Results after iteration number 9. First row: MSE versus rotation angle bins for the proposed method. Second row: number of samples selected in each angular bin for the proposed method. Best viewed in color

4.6 *Study of Number of Active Learning Iterations*

The purpose of this experiment was to study the effect of the number of active learning iterations on the performance of the proposed framework. We conducted experiments on the *WIKI Age Estimation Dataset* for $T = [5, 10, 15, 20, 25]$ and the mean squared error results are presented in Table 3. We observe that our framework depicts the least error compared to the baseline methods, across all the 5 values of T. This shows the robustness of our method to the number of active learning iterations. We further note that there is little performance gain after iteration number $T = 15$ while using all methods for this dataset. In a real-world setting, the number of iterations will be governed by the available labeling budget and will increase with an increase in budget.

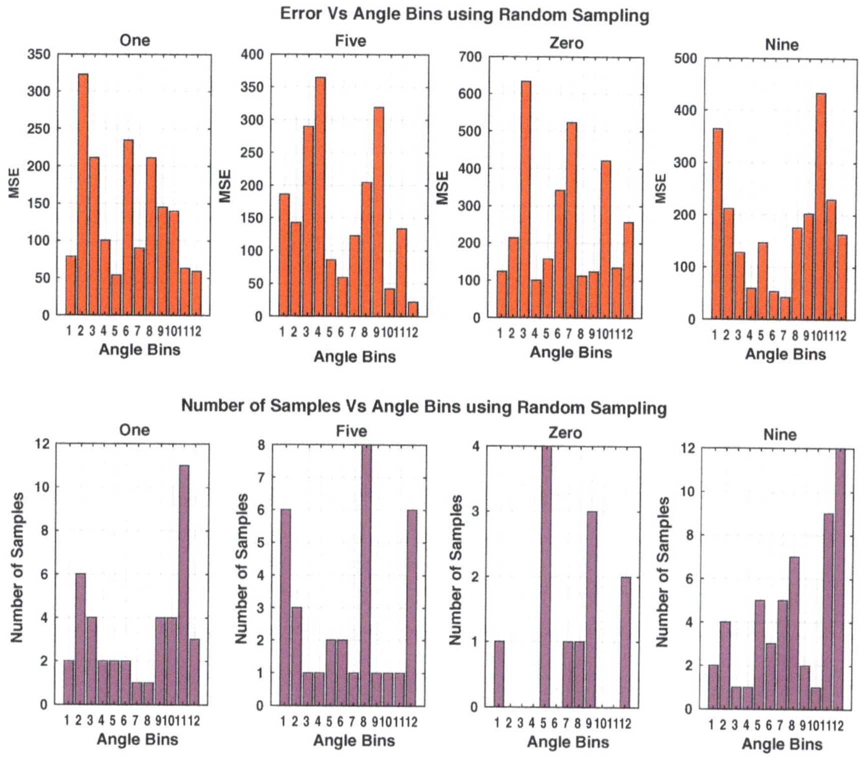

Fig. 10 Results after iteration number 9. First row: MSE versus rotation angle bins for random sampling. Second row: Number of samples selected in each angular bin for random sampling. Best viewed in color

Table 3 Study of a number of active learning iterations. The proposed framework shows the best performance at any given iteration when compared to all the baseline methods

Iteration number	Proposed method	Käding [23]	Greedy	QBC	Random sampling
5	87.5278	93.8099	121.3939	114.3875	137.8334
10	7.6926	20.1646	21.2742	22.7434	37.6184
15	5.0741	9.2519	10.3666	11.1212	19.0762
20	4.9965	8.2362	9.9948	9.4571	15.3716
25	4.2163	7.8990	9.0002	9.1188	14.9521

5 Conclusion and Future Work

In this paper, we proposed a novel deep active learning framework for regression applications. We used the Expected Model Output Change (EMOC) as the active selection criterion and integrated it within the objective function used to train the deep

CNN. The resulting model optimized this novel objective function and learned from salient examples that caused the maximum change to the current model. Extensive empirical studies on benchmark regression datasets (from a variety of application domains) demonstrated the effectiveness of the proposed framework in selecting the most informative samples for learning and annotation. Our in-depth analysis of the proposed active sampling criterion further corroborated the efficacy of our algorithm. To the best of our knowledge, this is the first research effort to leverage the feature learning capabilities of deep CNNs to develop a novel active learning algorithm for regression applications.

The proposed method uses the EMOC principle for quantifying the importance of an unlabeled sample. As explained in Sect. 3.2, a direct implementation of the EMOC principle would be computationally intensive; we, therefore, employed [23] and estimated the model change as the gradient of the objective with respect to a candidate example. The joint objective function used to train the deep CNN for active learning involves a higher training time as compared to the baseline methods used in this research. However, as seen from the results, our method depicts the best performance across all datasets and has the least error among all competing methods. Further, as seen from Table 2, our framework requires the least number of labels to achieve a given performance level. Thus, there is a trade-off between the computational complexity (time taken to train the model) and the performance of the algorithm. Depending on the available budget, time and computational resources in a real-world application, an appropriate algorithm needs to be selected.

As part of future work, we plan to study the effects of various parameters (such as the batch size k) on the active learning performance. We also intend to study the performance of our deep active learning framework on applications other than regression, such as data involving multi-labels and hierarchical labels.

References

1. J. Azimi, A. Fern, X. Zhang-Fern, G. Borradaile, B. Heeringa, Batch active learning via coordinated matching (2012), arXiv:1206.6458
2. T. Baltrušaitis, P. Robinson, L.P. Morency, 3D constrained local model for rigid and non-rigid facial tracking, in *2012 IEEE Conference on Computer Vision and Pattern Recognition (CVPR)* (IEEE, 2012), pp. 2610–2617
3. V. Belagiannis, S. Amin, M. Andriluka, B. Schiele, N. Navab, S. Ilic, 3D pictorial structures for multiple human pose estimation, in *Proceedings of the IEEE Conference on Computer Vision and Pattern Recognition* (2014), pp. 1669–1676
4. K. Brinker, Incorporating diversity in active learning with support vector machines, in *Proceedings of the 20th International Conference on Machine Learning (ICML-03)* (2003), pp. 59–66
5. R. Burbidge, J.J. Rowland, R.D. King, Active learning for regression based on query by committee, in *International Conference on Intelligent Data Engineering and Automated Learning* (Springer, 2007), pp. 209–218
6. W. Cai, Y. Zhang, J. Zhou, Maximizing expected model change for active learning in regression, in *2013 IEEE 13th International Conference on Data Mining (ICDM)* (IEEE, 2013), pp. 51–60

7. P. Campigotto, A. Passerini, R. Battiti, Active learning of pareto fronts. IEEE Trans. Neural Netw. Learn. Syst. **25**(3), 506–519 (2014)
8. S. Chakraborty, V. Balasubramanian, S. Panchanathan, Adaptive batch mode active learning. IEEE Trans. Neural Netw. Learn. Syst. **26**(8), 1747–1760 (2015)
9. R. Chattopadhyay, Z. Wang, W. Fan, I. Davidson, S. Panchanathan, J. Ye, Batch mode active sampling based on marginal probability distribution matching. ACM Trans. Knowl. Discov. Data (TKDD) **7**(3), 13 (2013)
10. D.A. Cohn, Z. Ghahramani, M.I. Jordan, Active learning with statistical models. J. Artif. Intell. Res. (1996)
11. A. Dosovitskiy, J.T. Springenberg, T. Brox, Learning to generate chairs with convolutional neural networks, in *2015 IEEE Conference on Computer Vision and Pattern Recognition (CVPR)* (IEEE, 2015), pp. 1538–1546
12. D. Eigen, C. Puhrsch, R. Fergus, Depth map prediction from a single image using a multi-scale deep network, in *Advances in neural information processing systems* (2014), pp. 2366–2374
13. E.J. de Fortuny, D. Martens, Active learning-based pedagogical rule extraction. IEEE Trans. Neural Netw. Learn. Syst. **26**(11), 2664–2677 (2015)
14. Y. Freund, H.S. Seung, E. Shamir, N. Tishby, Selective sampling using the query by committee algorithm. Mach. Learn. **28**(2–3), 133–168 (1997)
15. A. Freytag, E. Rodner, J. Denzler, Selecting influential examples: Active learning with expected model output changes, in *European Conference on Computer Vision* (Springer, 2014), pp. 562–577
16. R. Girshick, J. Donahue, T. Darrell, J. Malik, Rich feature hierarchies for accurate object detection and semantic segmentation, in *Proceedings of the IEEE Conference on Computer Vision and Pattern Recognition* (2014), pp. 580–587
17. G. Gkioxari, B. Hariharan, R. Girshick, J. Malik, R-CNNS for pose estimation and action detection (2014), arXiv:1406.5212
18. Y. Guo, Active instance sampling via matrix partition, in *Advances in Neural Information Processing Systems* (2010), pp. 802–810
19. Y. Guo, D. Schuurmans, Discriminative batch mode active learning, in *Advances in Neural Information Processing Systems* (2008), pp. 593–600
20. S.C. Hoi, R. Jin, M.R. Lyu, Large-scale text categorization by batch mode active learning, in *Proceedings of the 15th International Conference on World Wide Web* (ACM, 2006), pp. 633–642
21. S.C. Hoi, R. Jin, M.R. Lyu, Batch mode active learning with applications to text categorization and image retrieval. IEEE Trans. Knowl. Data Eng. **21**(9), 1233–1248 (2009)
22. M. Jaderberg, K. Simonyan, A. Vedaldi, A. Zisserman, Reading text in the wild with convolutional neural networks. Int. J. Comput. Vis. **116**(1), 1–20 (2016)
23. C. Käding, E. Rodner, A. Freytag, J. Denzler, Active and continuous exploration with deep neural networks and expected model output changes (2016), arXiv:1612.06129
24. A. Krizhevsky, I. Sutskever, G.E. Hinton, Imagenet classification with deep convolutional neural networks, in *Advances in Neural Information Processing Systems* (2012), pp. 1097–1105
25. D. Laptev, N. Savinov, J.M. Buhmann, M. Pollefeys, Ti-pooling: transformation-invariant pooling for feature learning in convolutional neural networks, in *Proceedings of the IEEE Conference on Computer Vision and Pattern Recognition* (2016), pp. 289–297
26. Y. LeCun, Y. Bengio, G. Hinton, Deep learning. Nature **521**(7553), 436 (2015)
27. D.D. Lewis, W.A. Gale, A sequential algorithm for training text classifiers, in *Proceedings of the 17th Annual International ACM SIGIR Conference on Research and Development in Information Retrieval* (Springer, New York, 1994), pp. 3–12
28. Z. Liu, X. Li, P. Luo, C.C. Loy, X. Tang, Semantic image segmentation via deep parsing network, in *2015 IEEE International Conference on Computer Vision (ICCV)* (IEEE, 2015), pp. 1377–1385
29. S. Liu, Y. Zhao, F. Xue, B. Chen, X. Chen, Deepcount: crowd counting with wifi via deep learning (2019), arXiv:1903.05316

30. T. Pfister, K. Simonyan, J. Charles, A. Zisserman, Deep convolutional neural networks for efficient pose estimation in gesture videos, in *Asian Conference on Computer Vision* (Springer, 2014), pp. 538–552
31. H. Ranganathan, S. Chakraborty, S. Panchanathan, Multimodal emotion recognition using deep learning architectures, in *IEEE Winter Conference on Applications of Computer Vision (WACV)* (2016)
32. H. Ranganathan, S. Chakraborty, S. Panchanathan, Transfer of multimodal emotion features in deep belief networks, in *2016 50th Asilomar Conference on Signals, Systems and Computers* (IEEE, 2016), pp. 449–453
33. H. Ranganathan, H. Venkateswara, S. Chakraborty, S. Panchanathan, Deep active learning for image classification, in *IEEE International Conference on Image Processing (ICIP)* (2017)
34. H. Ranganathan, H. Venkateswara, S. Chakraborty, S. Panchanathan, Multi-label deep active learning with label correlation, in *IEEE International Conference on Image Processing (ICIP)* (2018)
35. R. Ranjan, V.M. Patel, R. Chellappa, Hyperface: A deep multi-task learning framework for face detection, landmark localization, pose estimation, and gender recognition. IEEE Trans. Pattern Anal. Mach. Intell. (2017)
36. R. Rothe, R. Timofte, L. Van Gool, Dex: deep expectation of apparent age from a single image, in *IEEE International Conference on Computer Vision Workshops* (2015), pp. 10–15
37. R. Rothe, R. Timofte, L. Van Gool, Deep expectation of real and apparent age from a single image without facial landmarks. Int. J. Comput. Vis. **126**(2–4), 144–157 (2018)
38. B. Settles, M. Craven, An analysis of active learning strategies for sequence labeling tasks, in *Proceedings of the Conference on Empirical Methods in Natural Language Processing* (Association for Computational Linguistics, 2008), pp. 1070–1079
39. J. Sherrah, S. Gong, Fusion of perceptual cues for robust tracking of head pose and position. Pattern Recognit. **34**(8), 1565–1572 (2001)
40. Z. Shi, L. Zhang, Y. Liu, X. Cao, Crowd counting with deep negative correlation learning, in *IEEE Conference on Computer Vision and Pattern Recognition (CVPR)* (2018)
41. F. Stark, C. Hazırbas, R. Triebel, D. Cremers, Captcha recognition with active deep learning, in *Workshop New Challenges in Neural Computation 2015* (Citeseer, 2015), p. 94
42. M. Sugiyama, Active learning in approximately linear regression based on conditional expectation of generalization error. J. Mach. Learn. Res. **7**, 141–166 (2006)
43. M. Sugiyama, S. Nakajima, Pool-based active learning in approximate linear regression. Mach. Learn. **75**(3), 249–274 (2009)
44. Y. Sun, X. Wang, X. Tang, Deep convolutional network cascade for facial point detection, in *2013 IEEE Conference on Computer Vision and Pattern Recognition (CVPR)* (IEEE, 2013), pp. 3476–3483
45. C. Szegedy, A. Toshev, D. Erhan, Deep neural networks for object detection, in *Advances in Neural Information Processing Systems* (2013), pp. 2553–2561
46. S. Tong, D. Koller, Support vector machine active learning with applications to text classification. J. Mach. Learn. Res. **2**, 45–66 (2001)
47. A. Toshev, C. Szegedy, Deeppose: human pose estimation via deep neural networks, in *Proceedings of the IEEE Conference on Computer Vision and Pattern Recognition* (2014), pp. 1653–1660
48. D. Wang, Y. Shang, A new active labeling method for deep learning, in *2014 International Joint Conference on Neural Networks (IJCNN)* (IEEE, 2014), pp. 112–119
49. X. Wang, R. Guo, C. Kambhamettu, Deeply-learned feature for age estimation, in *IEEE Winter Conference on Applications of Computer Vision (WACV)* (2015)
50. X. Wang, L. Zhang, L. Lin, Z. Liang, W. Zuo, Deep joint task learning for generic object extraction, in *Advances in Neural Information Processing Systems* (2014), pp. 523–531
51. M.A. Wani, F.A. Bhat, S. Afzal, A.I. Khan, *Advances in Deep Learning*, vol. 57. (Springer, Berlin, 2020)
52. R. Willett, R. Nowak, R.M. Castro, Faster rates in regression via active learning, in *Advances in Neural Information Processing Systems* (2006), pp. 179–186

53. H. Yu, S. Kim, Passive sampling for regression, in *2010 IEEE 10th International Conference on Data Mining (ICDM)* (IEEE, 2010), pp. 1151–1156
54. S. Zaghbani, N. Boujneh, M. Bouhlel, Age estimation using deep learning. Comput. Electr. Eng. **68**, 1337–1347 (2018)
55. Z. Zhang, P. Luo, C.C. Loy, X. Tang, Facial landmark detection by deep multi-task learning, in *European Conference on Computer Vision* (Springer, 2014), pp. 94–108
56. C. Zhang, H. Li, X. Wang, X. Yang, Cross-scene crowd counting via deep convolutional neural networks, in *IEEE Conference on Computer Vision and Pattern Recognition (CVPR)* (2015)
57. Z. Zhao, H. Li, R. Zhao, X. Wang, Crossing-line crowd counting with two-phase deep neural networks, in *European Conference on Computer Vision (ECCV)* (2016)
58. S. Zhou, Q. Chen, X. Wang, Active deep networks for semi-supervised sentiment classification, in *Proceedings of the 23rd International Conference on Computational Linguistics: Posters* (Association for Computational Linguistics, 2010), pp. 1515–1523
59. X. Zhu, J. Lafferty, Z. Ghahramani, Combining active learning and semi-supervised learning using Gaussian fields and harmonic functions, in *ICML 2003 Workshop on the Continuum from Labeled to Unlabeled Data in Machine Learning and Data Mining*, vol. 3 (2003)

Deep Learning Based Hazard Label Object Detection for Lithium-ion Batteries Using Synthetic and Real Data

Anneliese Schweigert, Christian Blesing and Christoph M. Friedrich

Abstract Changing transport regulations for intralogistics tasks leads to the need for object detection of hazard labels on parcels in high-resolution grayscale images. For this reason, this paper compares different Convolutional Neural Network (CNN) based object detection systems. Specifically, a YOLO implementation known as Darkflow as well as a self-developed Object Detection Pipeline (ODP) based on the Inception V3 model is considered. Different datasets consisting of synthetic and real images are created to set up the necessary training and evaluation environments. To check the robustness of the systems under real operation conditions, they are assessed by the mean Average Precision (mAP) metric. Moreover, results are evaluated to answer various questions like the impact of synthetic data during training, or the highest quality level of the systems. The YOLO models showed a higher mAP and a much higher detection speed than the MSER Object Detection Pipeline at the cost of higher training times. The mixed training data set with synthetic and real data showed a slightly reduced mAP on the validation set compared to just real data.

Keywords Object detection · Synthetic data · Real data · Deep learning · Hazard label detection

A. Schweigert · C. Blesing
Fraunhofer Institute for Material Flow and Logistics (IML), Dortmund, Germany
e-mail: anneliese.schweigert@iml.fraunhofer.de

C. Blesing
e-mail: christian.blesing@iml.fraunhofer.de

A. Schweigert · C. M. Friedrich (✉)
Department of Computer Science, University of Applied Sciences and Arts Dortmund, Dortmund, Germany
e-mail: christoph.friedrich@fh-dortmund.de

M. A. Wani et al. (eds.), *Deep Learning Applications*,
Advances in Intelligent Systems and Computing 1098,
https://doi.org/10.1007/978-981-15-1816-4_8

1 Introduction

Nowadays, a lot of different traffic carriers like air, road, water, and rail are used to transport dangerous or hazardous goods, such as lithium-ion batteries. Inadequate transport of parcels with these goods inside, can cause accidents or lead to explosions and fires [23]. Companies of the courier, express, and parcel (CEP) business area, like FedEx and DHL, offer the service to transport a parcel from point A to point B. As the prevention of accidents, these companies need to adapt to continuously changing regulations. These regulations apply especially to the transport of lithium-ion batteries by air and are set by the International Civil Aviation Organization (ICAO) and the International Air Transport Association (IATA) [14]. As a prerequisite, the shipper has to support the attachment of the standardized labels for the transport of lithium-ion batteries (see Fig. 1). The courier has to ensure, that the labeled parcel which contains dangerous materials is transported properly and separately.

Despite the standardization of hazard labels, there are still problems with the attachment of the labels. For instance, some shippers attach two or more labels on a parcel. Alternatively, they print the label on the parcel or use a different label combination than the standardized. Another difficulty is that the hazard labels may get polluted by environmental influences during transport. All these aspects complicate the machine recognition of the symbols.

In this paper, object detection systems using deep learning or image processing techniques are selected to detect all symbols on the parcels.

This paper is organized as follows. At first, an overview of related works is given. Then the applied object detection systems are introduced in Sect. 3. The creation and application of the used training, validation, and test sets are explained in Sect. 4. The performance metric used is specified in Sect. 5 and the underlying CNN models are presented in Sect. 6. In Sect. 7, the training of the systems is described and in Sect. 8, the systems are evaluated and critically examined. Finally, conclusions are drawn.

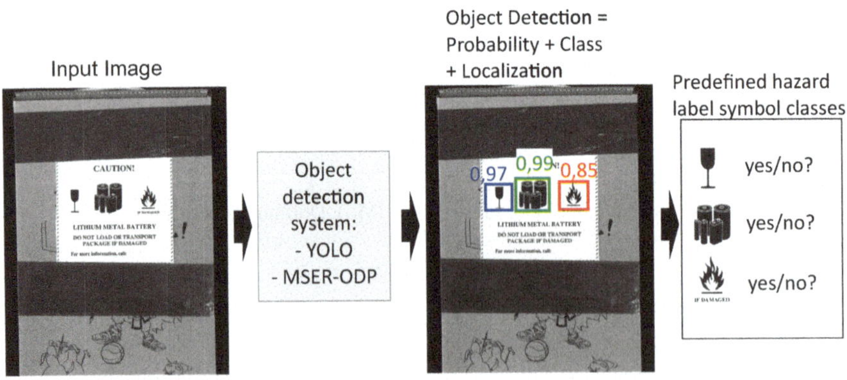

Fig. 1 The illustration depicts the use case. High-resolution grayscale images were taken by a 2D camera system showing parcels with labels. The images are used as input data for the different object detection systems. Finally, it can be derived, how parcels have to be transported correctly

2 Related Work

Nowadays, Convolutional Neuronal Networks (CNN) are most commonly applied in the area of object detection [4] and text recognition [21] in images. Moreover, they are used for audiovisual speech recognition [22] or video-based emotion [6] and action recognition [13].

In addition to these primary areas of application, CNN technology is constantly encountering new areas of application, particularly in mobile robotics. Especially in the context of global path planning [25], mapping and real- time semantic feature extraction [12].

CNNs are increasingly being used in intralogistics tasks. Considering the picking process, there are areas of application such as human activity recognition during order picking [10] or product picking by a stationary manipulator [33]. Furthermore, as maintenance and repair assistance in industrial inspection using Google Glass [26] represents another application area for CNNs.

2.1 *Synthetic Data*

Synthetic data is computer-generated data through data augmentation techniques [20] like blur, contrast, rotation, cropping, and much more. One of the biggest advantages of synthetic data is that it can be generated quickly and in very large amounts. In contrast to real data, the annotations can be created automatically during the data generation process.

Synthetic data is increasingly being used to train common CNN-based systems for text localization [11] and figure separation [35]. In the work of Jo et al. [15], the usage of synthetic and real data with Faster R-CNN [29] for object detection is compared. They showed that synthetic data alone produces a higher mAP than just real data and it is produced faster than real data. Synthetic-generated Captchas are used to train a CNN [16] to regress latent variables of a Captcha. Another application is the generation of synthetic data by a Generative Adversarial Network (GAN) [9, 20].

2.2 *Transfer Learning*

Transfer learning is commonly understood as learning a new task by using knowledge of a related already learned task [24]. Adapted to CNN's, a pretrained network with the knowledge of a training set is fine-tuned on a different training set.

To obtain a pretrained network usually, the ImageNet [31] training set is used. The fine-tuning step is classically used for the subsequent fine-tuning, e.g., classification

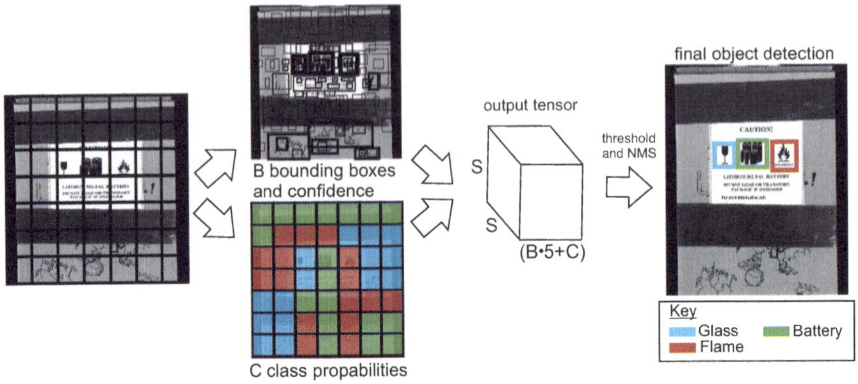

Fig. 2 The input is a realistic image. The number of anchors is represented by B and C is the number of classes. For each cell, B bounding boxes are calculated, visually depicted as black boxes. As a part of a bounding box, a confidence value is calculated, which is drawn with the line strength of the box. The highest class probability is determined for each cell. The YOLO algorithms output can be represented by a $S \times S \times (B \cdot 5 + C)$ tensor. These bounding boxes are filtered by a threshold and Non-Maximum Suppression (NMS). The remaining bounding boxes are visualized in the final object detection image [27]

of plants [17]. Another utilization of transfer learning is in speech recognition for cross-lingual communication [36] and in text categorization using Twitter tweets [3].

3 Methodology

Within this paper, two object detection systems are compared for hazard-label symbol detection. First, the original You Only Look Once (YOLO) algorithm [28] and second the self-developed Maximally Stable Extremal Regions Object Detection Pipeline (MSER-ODP).

3.1 YOLO

The YOLO algorithm [28] provides an end-to-end architecture (see Fig. 2) for object detection.

It splits the input image into a $S \times S$ grid. For each cell of the grid with respect to the number of anchors, B bounding boxes and C class probabilities are calculated. For every bounding box, five components are computed. The first four components represent the localization of the potential symbol in the input image and the fifth gives a confidence probability. The YOLO algorithm outputs visually depicted a 3D tensor containing the calculated bounding boxes and class probabilities for each

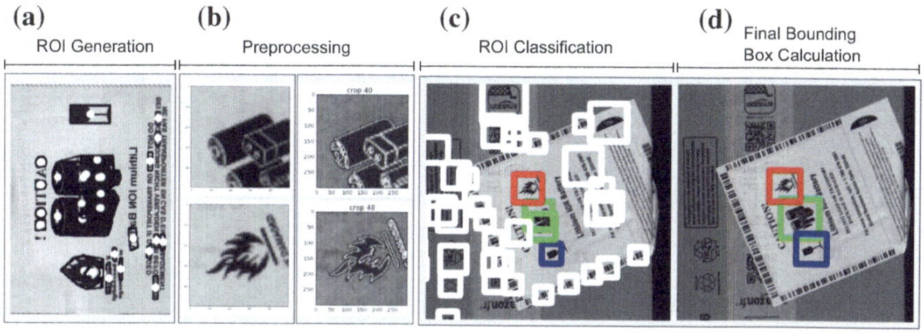

Fig. 3 This illustration shows the steps of the MSER-ODP. **a** The output of MSER are the blobs drawn in black. For each blob, the center of gravity is calculated and represented as white dots. (b-left side) During the Region of Interest (ROI) generation step, these blobs are cropped around their area of gravity. **b**-right side In another preprocessing step, they are manipulated by brightness, contrast, and adjusted to the expected size of the CNN model. **c** In the ROI classification step, they are classified through a CNN to four classes (Battery = green, Glass = blue, Flame = red, Background = white). As a result of the cropping, the position of the classified ROIs is known. They are visually depicted as boxes in different colors. **d** Unimportant crops containing background are discarded. The others are merged with regards to their class probabilities and positions

cell. These bounding boxes are filtered by a user-defined threshold. If more than one calculated bounding box is left, Non-Maximum Suppression (NMS) [30] is applied. For the computation of the bounding boxes and class probabilities, underlying CNN models like Tiny YOLO and YOLOv2 are used, which are presented in Sect. 6.1. For evaluation, the Darkflow[1] implementation of the YOLO algorithm [28] is used.

3.2 *MSER-ODP*

In order to detect the symbols in the input image, MSER-ODP uses four consecutive steps similar to the R-CNN approach [7]. These steps are depicted in Fig. 3 and the details of these steps are described in the following subsections.

The Object Detection Pipeline (MSER-ODP) uses the MSER algorithm [18] and the Inception V3 model [34] from the TF-slim library.[2] This CNN model is presented in Sect. 6.2.

[1]Darkflow, URL: https://github.com/thtrieu/darkflow (Retrieved: 2019-06-19).

[2]Nathan Silberman and Sergio Guadarrama, TensorFlow-Slim image classification model library, URL: https://github.com/tensorflow/models/tree/master/research/slim (Retrieved: 2019-06-19).

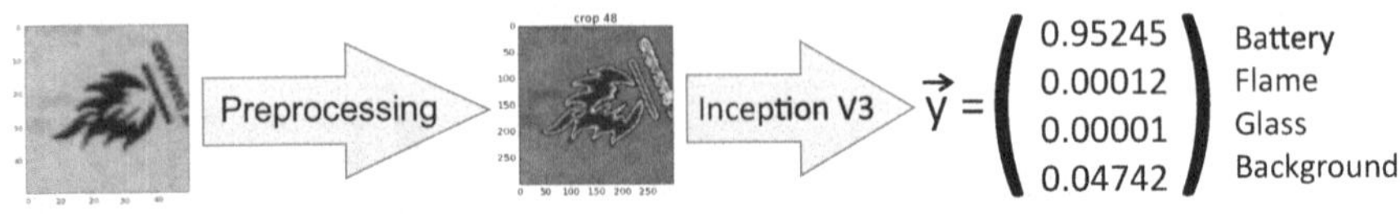

Fig. 4 This picture shows an illustrative input and output of the Inception V3 model

3.2.1 Region of Interest (ROI) Generation

The MSER-ODP approach is able to take an input image as shown in Fig. 1, which contains the parcel with the attached label and symbols. On this image, MSER [18] is applied, which outputs blobs. These blobs are used to extract ROIs at there center of gravity (see Fig. 3a). Every ROI extracted has quadratic dimension and variates in size (see Fig. 3c).

3.2.2 Preprocessing

After extraction, the ROIs must be preprocessed to be classified in the next step. Every ROI I_i is bilinearly interpolated to the fixed $299 \times 299 \times 3$ CNN input size and in addition, simultaneously normalized to the range of [0, 1]. First, the ROI I_i is darkened to half of the original brightness in every possible pixel position (x, y). The result is the image I_v (see Eq. 1).

$$I_v(x, y) = I_i(x, y) - 0.5 \tag{1}$$

ROI I_v is the input of the double contrast enhancement method, which outputs the image I_k (see Eq. 2). As a result, each pixel of it is normalized to $[-1, 1]$.

$$I_k(x, y) = I_v(x, y) \cdot 2.0 \tag{2}$$

An exemplary output of this introduced preprocessing is shown in Fig. 3b-right side. The edges of the symbols are enhanced.

3.2.3 ROI Classification

The extracted and resized ROIs from the previous step are classified by the Inception V3 [34] model. The output is customized to a 1D tensor with four elements/classes: Battery, Flame, Glass, and Background (see Fig. 4).

A detailed description of the Inception V3 training process and configuration is given in Sect. 7.1. As a result of the classification step, each ROI is assigned to one of the predefined classes using the softmax function. Additionally, the position of every ROI in the input image is known and represented by a bounding box (see Fig. 3c).

3.2.4 Calculation of the Final Bounding Boxes

In some cases, a symbol, such as the battery, may disintegrate into more than one ROI (see Fig. 3c). However, the desired result is calculating exactly one box for each of the three symbols (see Fig. 3d). To solve this problem, ROIs which are close to each other and are members of the same class were assigned to a group. All ROIs in a group are merged to form a final bounding box. Those groups belonging to the background class do not contain a searched symbol. They are excluded from the grouping/union process. Ultimately, the final bounding box of a group g is described as a 4-tupel $(xmin_g, xmax_g, ymin_g, ymax_g)$. For each group, the smallest $xmin$ and $ymin$ values and the largest $xmax$ and $ymax$ values over all n members are calculated (see Eq. 3 a-d):

$$xmin_g = min(\{xmin_0, \ \ldots, \ xmin_n\}) \tag{3a}$$

$$ymin_g = min(\{ymin_0, \ \ldots, \ ymin_n\}) \tag{3b}$$

$$xmax_g = max(\{xmax_0, \ \ldots, \ xmax_n\}) \tag{3c}$$

$$ymax_g = max(\{ymax_0, \ \ldots, \ ymax_n\}). \tag{3d}$$

4 Dataset Creation

In the following section, the used data sets are introduced. They are divided into datasets for YOLO and MSER-ODP. The overview of the set utilization is visualized in Fig. 5. For training, validation, and test, bounding box annotated real images have been manually annotated (see Sect. 4.1). The YOLO models are trained with the bounding box annotated training set. The MSER-ODP training set consists of single hazard label symbol images extracted from realistic parcel images. The first training set represents real hazard label symbols (see Sect. 4.2) and the second, mixed training set contains synthetic and real images (see Sect. 4.3).

All object detection systems are evaluated with the bounding box annotated validation and test sets. The validation set is used to optimize the training parameters. The test sets are independent of the training and validation sets and only seen once by the corresponding system. This described procedure enables a comparability of the systems.

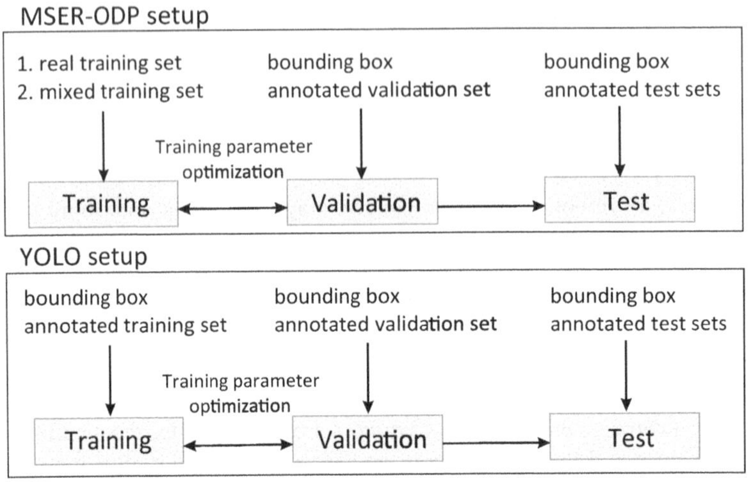

Fig. 5 The purpose of the different sets is visualized for different modeling phases (e.g., training, validation, and test)

Table 1 The bounding box training set is derived from 1991 images. The validation set is derived from 498 images. The test set is derived from 260 images

Class	Training set	Validation set	Test set	Sum
Battery	2010	505	269	2784
Flame	2009	505	267	2781
Glass	2011	505	268	2784
Sum	6030	1515	804	8349

4.1 Generation and Distribution of the Bounding Box Annotated Real Dataset

The original images taken by the camera system have a size of $6144/8192 \times H$ pixels. The variable H has different values because it depends on the depth of the specific parcel. Looking at all of these images, H is between 1696 and 11616 pixels. High-resolution images are a problem because they take a lot of computing resources. That is why they are resized to one-quarter of the original size with respect to the aspect ratios. The bounding box real dataset is manually annotated and consists of a training, validation, and test set (see Table 1).

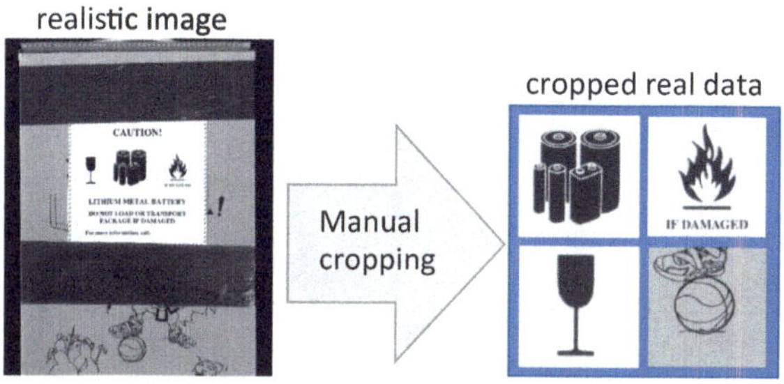

Fig. 6 This visualization shows the manual extraction of crops from a realistic image. Inside each of these 2749 images, every hazard label symbol is extracted as well as some crops for the background class. Sometimes more or less than four images are extracted depending on the labeling (see Sect. 1)

4.2 Generation and Distribution of the Real Training Set

For the MSER-ODP, single-label images are required to train the Inception V3 model (see Sect. 7.1). For this purpose, crops of realistic images recorded by the 2D camera system (see Fig. 1) are taken from the bounding box training images. Additionally, example images for the background class are manually annotated for the MSER-ODP real training set shown in Fig. 6.

From 2749 realistic images, 10349 single-label images are extracted. The dimension of these images is fixed to 200 $\times$ 200 pixels. Table 2 shows in the second column the distribution of the single-label training set consisting of cropped real data for each of the four predefined classes.

4.3 Generation and Distribution of the Mixed Training Set

A mixed training set with synthetic and real data is created to show the impact of synthetic data usage. The first component of this set is the cropped real images. The second component is synthetic data created through offline data augmentation.

Synthetic data is produced out of 8 original images of each hazard label symbol class and 6 original images of the background class. In total, 18027 images for the training set are produced by applying random distortions like duct tape and pen strokes to the original images. Random translations, scaling, and perspective distortion are also applied (see Fig. 7).

Each augmented image has the same size as the cropped real images, which means 200 $\times$ 200 pixels in grayscale. Table 2 shows the distribution of the mixed training set, which contains overall 26061 images over 4 classes. The same cropped images as in Sect. 4.2 are used for this training set.

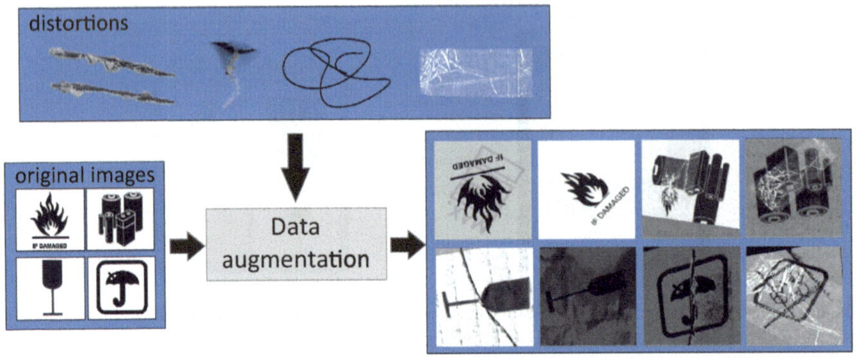

Fig. 7 For four classes (Battery, Flame, Glass, and Background) an offline data augmentation on original images is conducted. Each original image portrays just the symbol without distortions and transformations

Table 2 The distribution of the mixed MSER-ODP training set with synthetic and real data

Class	Training data (real)	Training data (synthetic)	Sum
Battery	2010	4011	6021
Flame	2009	4008	6017
Glass	2011	4008	6019
Background	2004	6000	8004
Sum	8034	18027	26061

5 Performance Metric

The mean Average Precision (mAP) metric was used in the PASCAL VOC Challenge 2007 [5]. Within this paper, this metric is used to assess the presented object detection systems.

Each predicted bounding box of a class i is classified into True Positive (TP), False Positive (FP), False Negative (FN) by means of Jaccard Index [2] with the threshold of 0.5 and the ground truth boxes of the validation or test set.

All predicted bounding boxes of a class i are sorted in descending order by there class confidence. From this the precision pr and recall re are calculated (see Eq. 4) on the accumulated number of TP and FP detections for each prediction. This is called precision–recall curve.

$$pr = \frac{TP}{TP + FP} \qquad re = \frac{TP}{TP + FN} \tag{4}$$

Each Average Precision AP_i of a class $i \in K$ is related to average the interpolated precisions p_{interp} over 11 equally distanced recall levels r. The interpolated precisions

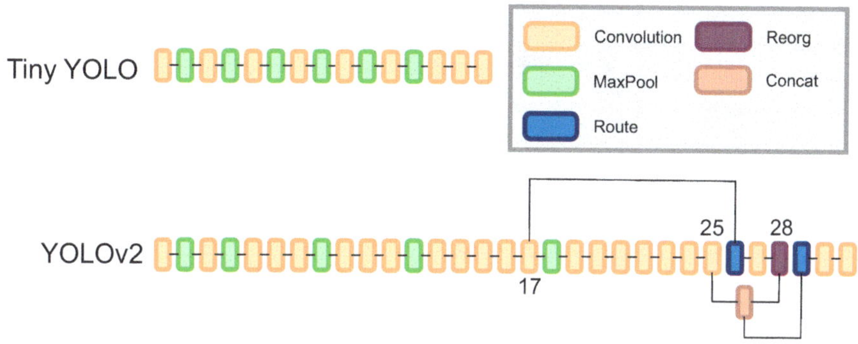

Fig. 8 This visualization of the YOLO models is based on the Darkflow source code

are obtained by taking the maximum precision whose recall re is greater than r from the precision–recall curve. The mAP is determined by the arithmetic mean over all AP_i (see Eq. 5).

$$AP_i = \frac{1}{11} \sum_{r \in \{0, 0.1, \ldots, 1\}} p_{interp}(r) \qquad mAP = \frac{1}{K} \sum_{i=1}^{K} AP_i \tag{5}$$

6 Architecture of the CNN Models

This section briefly explains architectural diagrams of the CNN models mentioned earlier. First, the YOLO models [28], and finally, the Inception V3 model [34] are explained.

6.1 Tiny YOLO and YOLOv2 Model Architectures

The Tiny YOLO model consists of 15 layers including 9 convolutional layers, and 6 max-pooling layers (see Fig. 8). The task of the convolutional layer is to extract learned features. The max-pooling layer is used for dimensionality reduction and overfitting prevention. Further details about these basic layers are not explained in this paper and [8] is recommended for details.

The YOLOv2 model contains 31 layers composed of 23 convolutional layers, 5 max-pooling layers, 2 route layers, and a reorg layer (see Fig. 8). The route layer feeds the output tensor of a previous layer into a deeper layer. An alternative is to feed equally sized output tensors from previous layers by means of a concatenation layer into a deeper layer. The reorg layer has the task to restructure a tensor.

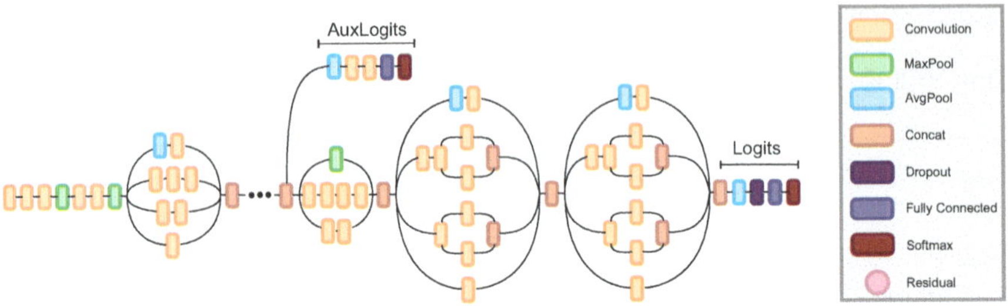

Fig. 9 This diagram visualizes the CNN architecture of the Inception V3 model. It is created on the foundation of TF-slim source code

6.2 *Inception V3 Model Architecture*

The structure of the Inception V3 is more complex than the architecture of the previously presented models, which are sequential oriented. The Inception V3 model executes the first seven layers and the AuxLogits and Logits area sequentially attached (see Fig. 9). The AuxLogits area is used during training of the network to reduce the effect of the vanishing gradient. Average pooling layers have a similar function as max- pooling layers, and are used downsampling to reduce overfitting. The concatenation layer chains output tensors of the same height and width. The dropout layer randomly disables the weights of a fully connected layer. The other areas are processed in parallel.

The chosen architecture has shown superior results in previous applications [1, 19]. Potentially other architectures as mentioned in [37] could further improve the results in this application.

7 Configuration and Training of the Object Detection Systems

Each CNN model (Inception V3, Tiny YOLO, and YOLOv2) is fine-tuned multiple times with different training parameter combinations. The results are the best training checkpoints with the highest mAP on the bounding box annotated validation set for each CNN model. In the end, the best runs are selected and presented in the following subsections.

For training of the object detection systems, an NVIDIA GTX 1080 with 8 GB vRAM and 2560 CUDA Cores is used. Additionally, the computer is equipped with an Intel(R) Xeon(R) CPU E5-2670 v2 @ 2.50 GHz with 10 Cores, 20 Threads, and 64 GB RAM.

Table 3 The table shows the training parameters and runtime of the CNN model Inception V3 of the MSER-ODP. At the 1. fine-tuning run, a checkpoint inception_v3 was provided by TF-slim. For the 2. fine-tuning run, the checkpoint of the first run is taken and the complete model is trained again with the same training data

System: MSER-ODP	1. Fine-tuning		2. Fine-tuning	
CNN model	Inception V3	Inception V3	Inception V3	Inception V3
Parameters	Training set			
	Mixed	Real	Mixed	Real
Checkpoint	inception_v3	inception_v3	3256	1004
Learning rate (lr)	0.01	0.01	0.0001	0.0001
lr decay	0.94	0.94	0.94	0.94
lr decay every x epochs	2	2	10	10
Optimizer	RMSProp	RMSProp	RMSProp	RMSProp
Momentum	–	–	–	–
Batch size	32	32	16	16
Steps	3256	1004	81400	12250
Runtime in dec. min./h	10.75 min	3.5 min	7.92 h	1.20 h

7.1 *Configuration and Training of the MSER-ODP*

Two different training sets are used to train the system as shown in Fig. 5. An Inception V3 model is trained with the real and mixed training set (see Table 2). The input volume is $299 \times 299 \times 3$ and the output volume is adjusted to four classes. For this, the two-phase transfer learning approach [32] with an ImageNet [31] pretrained model is used for training (see Table 3).

7.2 *Configuration and Training of the YOLO Approach*

To apply transfer learning (see Sect. 2.2) for object detection, CNN models denominated Tiny YOLO and YOLOv2 are used. These models are pretrained on PASCAL VOC data [5], which contains bounding box annotations for 20 different classes. In this work, the YOLO models are configured to fit the memory size of an NVIDIA GTX 1080 and three classes. The following explanation is based on Fig. 2. Tiny YOLO expects an input tensor of $1440 \times 1440 \times 3$. Consequentially, the network produces a $S \times S \times (B \cdot 5 + C)$ tensor with $S = 45$, $B = 5$ and $C = 3$. This results in 56700 variables for each image. The adjusted YOLOv2 model expects a $1248 \times 1248 \times 3$-tensor as input size. It converts an input image into a $S \times S \times (B \cdot 5 + C)$ tensor with $S = 39$, $B = 5$ and $C = 3$. Thus, the YOLOv2

Table 4 The training parameters and the resulting training duration of Tiny YOLO and YOLOv2 models with the Darkflow implementation can be seen. In the first run, pretrained checkpoints are used and in the second, the best checkpoint of the previous fine-tuning run

System: YOLO	1. Fine-tuning		2. Fine-tuning	
CNN model	Tiny YOLO	YOLOv2	Tiny YOLO	YOLOv2
Parameters	Training set			
	Real	Real	Real	Real
Checkpoint	Tiny-yolo-voc.weights	Yolo.weights	19500	33000
Learning rate (lr)	0.0001	0.0001	0.00001	0.00001
lr decay	–	–	–	–
lr decay every x epochs	–	–	–	–
Optimizer	ADAM	ADAM	ADAM	ADAM
Momentum	0.9	0.9	0.9	0.9
Batch size	4	2	4	2
Steps	19880	39800	2485	4075
Runtime in dec. min./h	16.92 h	21.46 h	1.99 h	2.56 h

model produces exactly 42588 variables per image. Furthermore, for each of these two models, the sizes of anchors and the number of applied filters of the last convolutional layer are adjusted. Both models have the same five anchors, which are quadratic and vary in size. The number of filters used at the last convolutional layer had to be adjusted by $B \cdot (C + 5)$ to get the correct output tensor volume.

Every input image is resized during the YOLO preprocessing into the input size of the corresponding CNN model.

The YOLO CNN models are fine-tuned with the bounding box annotated training set containing realistic images (see Table 1). Every time, all weights of the CNN models are trained (see Table 4). The first fine-tuning run adjusts the output size and the second optimizes the model with a smaller learning rate.

8 Questions and Results

After fine-tuning the models, the last checkpoint resulting from the second fine-tuning run of each model is used to answer the following questions.

How well is synthetically generated training data suitable for the identification of dangerous goods?

In our case, the enrichment of synthetic data has slightly reduced the mAP of the introduced MSER-ODP compared to training on real data alone by 4.4% (see

Table 5 The results on the bounding box annotated validation set with the Inception V3 model

System	CNN model	Training set	mAP on validation set (%)
MSER-ODP	Inception V3	Real	88.5
MSER-ODP	Inception V3	Mixed	84.1

Table 6 This table shows the results of each system on the bounding box annotated validation set

System	CNN model	Training set	mAP on validation set (%)
MSER-ODP	Inception V3	Real	88.5
YOLO	Tiny YOLO	Bounding box annotated	90.0
YOLO	YOLOv2	Bounding box annotated	90.0

Table 7 The evaluation on the bounding box annotated test set is depicted

System	CNN model	Training set	mAP on test set (%)
MSER-ODP	Inception V3	Real	88.6
YOLO	Tiny YOLO	Bounding box annotated	90.4
YOLO	YOLOv2	Bounding box annotated	90.3

Table 5). It can be concluded that synthetic training data is only conditionally suitable, which is explained by the limited variety of the synthetic images. This could be improved with more elaborated data augmentation techniques.

Which system has the highest quality level with the bounding box annotated validation set?
In Table 6, the results on the bounding box annotated validation set are presented. It shows that the Tiny YOLO and the YOLOv2 model have the highest mAP and thus the highest quality level. The MSER-ODP can compete well with the YOLO models.

How is the generalization performance of the systems with an independent test set?
The same checkpoints as in the previous question are used to determine the generalization performance of the systems. The results after fine-tuning the CNN models are shown in Table 7. It shows that the Tiny YOLO model is slightly better than the YOLOv2 model. The MSER-ODP can still compete well.

How the systems react with a test set with random contrast, brightness, and blur adjustments?

Table 8 The evaluation of the systems on the manipulated test set

System	CNN model	Training set	mAP on manipulated test set (%)
MSER-ODP	Inception V3	Real	73.1
YOLO	Tiny YOLO	Bounding box annotated	80.6
YOLO	YOLOv2	Bounding box annotated	87.0

The results in Table 8 were determined with the premises as the previous questions. On the unprocessed test images (see Table 2), different data augmentation methods are performed in a variety of combinations and sequences. It has been ensured that the pixels of the manipulated test images remain within the value range $\{0, \ldots, 255\}$. In brightness adjustment, a random number between -40 and 60 is selected. This random number is subtracted for each pixel of an image. A contrast adaption occurred in a range of 0.5–2.5. This value is also determined randomly and is the factor for each pixel of an image. The limits of the color range are also taken into account here. Blur is achieved with a 5×5 Gaussian filter.

Subsequently, an evaluation with the bounding box annotated test set with changes on the images is accomplished. As shown in the following table, YOLOv2 is more robust against random contrast, brightness and blur adjustments.

9 Conclusion

In this work, CNN-based object detection systems for hazard label detection are presented. Transfer learning is applied to these systems and they were used to answer the question if the detection could be improved when training data is enriched with synthetic data. The mixed training data set showed with the Inception V3 model a slightly reduced mAP by 4.4% on the validation set compared to just real data. Jo et al. [15], which have already been mentioned, concluded that synthetic data have an advantage in quantity and variety, but are not necessarily reflecting the real environment. Therefore, it might be concluded that there has to be more variety within the synthetic data to enrich the real data beneficially. Another question was if the newly introduced MSER-ODP outperforms the end-to-end object detection system YOLO on an independent test set. As a result, the YOLOv2 and Tiny YOLO models have each a higher mAP and a much higher detection speed than the MSER-ODP but the training time was higher. Finally, the robustness on the randomly manipulated test set has been assessed. The YOLO models showed superior results.

In the future, the data augmentation system to create synthetic data can be implemented with a GAN which might improve the variety of synthetic data. Additionally, other object detection systems can be investigated.

References

1. A. Canziani, A. Paszke, E. Culurciello, An analysis of deep neural network models for practical applications (2016)
2. S. Cesare, Y. Xiang, *Software Similarity and Classification* (Springer, Berlin, 2012), p. 67
3. A. Chachra, P. Mehndiratta, M. Gupta, Sentiment analysis of text using deep convolution neural networks, in *10th International Conference on Contemporary Computing (IC3)* (2017), pp. 1–6
4. Y.H. Chang, P.L. Chung, H.W. Lin, Deep learning for object identification in ROS-based mobile robots, in *IEEE International Conference on Applied System Invention (ICASI)* (2018), pp. 66–69
5. M. Everingham, S.M.A. Eslami, L.V. Gool, C.K.I. Williams, J. Winn, A. Zisserman, The pascal visual object classes challenge: a retrospective. Int. J. Comput. Vis. (IJCV) **111**(1), 98–136 (2015)
6. Y. Fan, X. Lu, D. Li, Y. Liu, Video-based emotion recognition using CNN-RNN and C3D hybrid networks, in *18th ACM International Conference on Multimodal Interaction, ICMI '16* (2016), pp. 445–450
7. R. Girshick, J. Donahue, T. Darrell, J. Malik, Rich feature hierarchies for accurate object detection and semantic segmentation, in *IEEE Conference on Computer Vision and Pattern Recognition (CVPR)* (2014), pp. 580–587
8. I. Goodfellow, Y. Bengio, A. Courville, *Deep Learning* (MIT Press, Cambridge, 2016)
9. I. Goodfellow, J. Pouget-Abadie, M. Mirza, B. Xu, D. Warde-Farley, S. Ozair, A. Courville, Y. Bengio, Generative adversarial nets. Adv. Neural Inf. Process. Syst. **27**, 2672–2680 (2014)
10. R. Grzeszick, J.M. Lenk, F. Moya Rueda, G.A. Fink, S. Feldhorst, M. ten Hompel, Deep neural network based human activity recognition for the order picking process, in *4th International Workshop on Sensor-based Activity Recognition and Interaction, iWOAR '17* (2017) pp. 1–6
11. A. Gupta, A. Vedaldi, A. Zisserman, Synthetic data for text localisation in natural images, in *IEEE Conference on Computer Vision and Pattern Recognition (CVPR)* (2016), pp. 2315–2324
12. M. Himstedt, E. Maehle, Online semantic mapping of logistic environments using RGB-D cameras. Int. J. Adv. Robot. Syst. **14**(4), 113 (2017)
13. R. Hou, C. Chen, M. Shah. Tube convolutional neural network (T-CNN) for action detection in videos, in *IEEE International Conference on Computer Vision (ICCV)* (2017), pp. 5823–5832
14. IATA DGR, 2017 Lithium Battery Guidance Document. https://www.iata.org/whatwedo/cargo/dgr/Documents/lithium-battery-guidance-document-2017-en.pdf (2016) (Retrieved: 2019-06-19)
15. H. Jo, Y. H. Na, J.B. Song, Data augmentation using synthesized images for object detection, in *17th International Conference on Control, Automation and Systems (ICCAS)* (2017), pp. 1035–1038
16. T.A. Le, A.G. Baydin, R. Zinkov, F. Wood, Using synthetic data to train neural networks is model-based reasoning, in *International Joint Conference on Neural Networks (IJCNN)* (2017), pp. 3514–3521
17. A.R. Ludwig, H. Piorek, A.H. Kelch, D. Rex, S. Koitka, C.M. Friedrich, Improving model performance for plant image classification with filtered noisy images, in *Working Notes of CLEF 2017 - Conference and Labs of the Evaluation Forum*, vol. 1866 (2017)
18. J. Matas, O. Chum, M. Urban, T. Pajdla, Robust wide baseline stereo from maximally stable extremal regions, in *British Machine Vision Conference (BMVC 2002)*, vol. 22 (BMVA Press, 2002), pp. 384–393
19. Y. Mednikov, S. Nehemia, B. Zheng, O. Benzaquen, D. Lederman, Transfer representation learning using inception-V3 for the detection of masses in mammography, in *40th Annual International Conference of the IEEE Engineering in Medicine and Biology Society (EMBC)* (2018), pp. 2587–2590
20. A. Mikołajczyk, M. Grochowski, Data augmentation for improving deep learning in image classification problem, in *International Interdisciplinary PhD Workshop (IIPhDW)* (2018), pp. 117–122

21. Y. Nagaoka, T. Miyazaki, Y. Sugaya, S. Omachi, Text detection by faster R-CNN with multiple region proposal networks, in *14th IAPR International Conference on Document Analysis and Recognition (ICDAR)*, vol. 06 (2017), pp. 15–20
22. K. Noda, Y. Yamaguchi, K. Nakadai, Hiroshi G. Okuno, T. Ogata, Audio-visual speech recognition using deep learning. Appl. Intell. **42**(4), 722–737 (2015)
23. G. Nowacki, C. Krysiuk, R. Kopczewski, Dangerous goods transport problems in the European Union and Poland. TransNav, Int. J. Mar. Navig. Saf. Sea Transp. **10**(1), 143–150 (2016)
24. S.J. Pan, Q. Yang, A survey on transfer learning. IEEE Trans. Knowl. Data Eng. **22**(10), 1345–1359 (2010)
25. D. Pande, C. Sharma, V. Upadhyaya, Object detection and path finding using monocular vision, in *International Conference on Signal Propagation and Computer Technology (ICSPCT)* (2014), pp. 376–379
26. P. Ramakrishna, E. Hassan, R. Hebbalaguppe, M. Sharma, G. Gupta, L. Vig, G. Sharma, G. Shroff, An ar inspection framework: feasibility study with multiple ar devices, in *IEEE International Symposium on Mixed and Augmented Reality (ISMAR-Adjunct)* (2016), pp. 221–226
27. J. Redmon, S. Divvala, R. Girshick, A. Farhadi, You only look once: unified, real-time object detection, in *IEEE Conference on Computer Vision and Pattern Recognition (CVPR)* (2016), pp. 779–788
28. J. Redmon, A. Farhadi, YOLO9000: better, faster, stronger, in *IEEE Conference on Computer Vision and Pattern Recognition (CVPR)* (2017), pp. 6517–6525
29. S. Ren, K. He, R. Girshick, J. Sun, Faster R-CNN: towards real-time object detection with region proposal networks, in *Advances in Neural Information Processing Systems 28 (NIPS)* (2015), pp. 91–99
30. R. Rothe, M. Guillaumin, L. Van Gool, Non-maximum suppression for object detection by passing messages between windows, in *Computer Vision – ACCV 2014* (Springer International Publishing, 2015), pp. 290–306
31. O. Russakovsky, J. Deng, H. Su, J. Krause, S. Satheesh, S. Ma, Z. Huang, A. Karpathy, A. Khosla, M. Bernstein, A.C. Berg, F. Li, ImageNet large scale visual recognition challenge. Int. J. Comput. Vis. (IJCV) **115**(3), 211–252 (2015)
32. Y. Sawada, Y. Sato, T. Nakada, S. Yamaguchi, K. Ujimoto, N. Hayashi, Improvement in classification performance based on target vector modification for all-transfer deep learning. Appl. Sci. **9**(1) (2019)
33. M. Schwarze, A. Milan, A.S. Periyasamy, S. Behnke, RGB-D object detection and semantic segmentation for autonomous manipulation in clutter. Int. J. Robot. Res. **37**(4–5), 437–451 (2017)
34. C. Szegedy, V. Vanhoucke, S. Ioffe, J. Shlens, Z. Wojna, Rethinking the inception architecture for computer vision, in *IEEE Conference on Computer Vision and Pattern Recognition (CVPR)* (2016), pp. 2818–2826
35. S. Tsutsui, D. Crandall, A data driven approach for compound figure separation using convolutional neural networks, in *The IAPR International Conference on Document Analysis and Recognition (ICDAR)* (2017)
36. J. Wang, S. Liu, Q. Yang, Transfer learning for air traffic control LVCSR system, in *Second International Conference on Mechanical, Control and Computer Engineering (ICMCCE)* (2017), pp. 169–172
37. M.A. Wani, F.A. Bhat, S. Afzal, A. Khan, *Advances in Deep Learning* (Springer, Berlin, 2020)

Enabling Robust and Autonomous Materialhandling in Logistics Through Applied Deep Learning Algorithms

Christian Poss, Thomas Irrenhauser, Marco Prueglmeier, Daniel Goehring, Vahid Salehi and Firas Zoghlami

Abstract In recent years, logistics costs in the automotive industry have risen significantly. One way to reduce these costs is to automate the entire material flow. To meet the flexible industrial challenges and dynamic changes, robots with intelligent perception are necessary. Such a perception algorithm is presented in the following. It consists of three modules. In the first module, all objects in the field of vision of the robot are detected, and their position is determined. Then the relevant objects for the respective process are selected. Finally, the gripping point of the next object to be handled is determined. By integrating the robots, it can be shown that by combining intelligent modules with pragmatic frame modules, automation in a challenging industrial environment is feasible.

1 Introduction

In recent years, logistics has developed more and more away from a real enabler to vehicle production toward a core component of it. This is reflected, for example, in

C. Poss (✉) · T. Irrenhauser · M. Prueglmeier
BMW Group, Munich, Germany
e-mail: christian.poss@bmw.de

T. Irrenhauser
e-mail: thomas.irrenhauser@bmwgroup.com

M. Prueglmeier
e-mail: marco.prueglmeier@bmw.de

D. Goehring
Freie Universiterlins Berlin, Berlin, Germany
e-mail: daniel.goehring@fu-berlin.de

V. Salehi · F. Zoghlami
University of Applied Sciences, Munich, Germany
e-mail: vahid.salehi_douzloo@hm.edu

F. Zoghlami
e-mail: firas.zoghlami@hm.edu

M. A. Wani et al. (eds.), *Deep Learning Applications*,
Advances in Intelligent Systems and Computing 1098,
https://doi.org/10.1007/978-981-15-1816-4_9

the share of logistics costs per part. Logistics activities cause almost 25% of the total cost of manufacturer purchase prices. The remaining 75% are mainly composed of material costs (66%) and production costs (15%).

This increase in costs is mainly due to the increased complexity of the entire logistics system. The critical contributing factors are the increase of the variety of offered vehicle models as well as the respective individualization possibilities of these vehicle derivatives. For example, the number of vehicle models in the BMW Group rose from 6 (1990) to 29 within 25 years. Due to the personalization possibilities of these vehicles demanded by customers, there are now 10 to the power of 32 possible variants when ordering a car. Together with the reduced vertical range of manufacture due to advancing globalization and increased cost pressure, these developments have led to a significant increase in the number of components to be delivered to the assembly line in the vehicle plants. While in 2000 more than half (55%) of the body components were still produced by the OEM themselves, this figure fell to 29% by 2015.

The increase in the number of parts to be provided leads directly to a rise in the work steps required in the plants, which ultimately results in additional personnel requirements. Due to the growth efforts of the automotive OEM as such, the number of vehicles to be produced in the individual plants was continuously increased in parallel. This leads to a further strain on the personnel situation, especially in those areas where the relaxed labor market situation has already led to almost full employment.

One practical approach that encourages productivity while reducing the costs is the holistic automation of the logistics material flow in the plants. In addition to the economic gains, it reduces the need for human employment in monotonous and unergonomic handling tasks and promotes more demanding working areas such as final vehicle assembly or quality assurance.

1.1 Evolving Automatisation

The high-dynamic production processes in the modern industry increase the complexity of logistics and thus encourages the growing adoption of automation solutions. Therefore, robotic solutions and intelligent machines are transforming manufacturing industries to a higher level of automation. The number of industrial robots worldwide will increase from 1.828.000 (2016) to over 3 million by 2020. The main markets are Asian countries such as China, South Korea, and Japan as well as America and Germany.

With falling costs and innovative business models such as robot leasing, this trend is affecting not only the production halls of large OEM but also those of small and medium-sized enterprises.

The highest use can be observed in the electronics industry, in particular, in the automation of the thoroughly repetitive production steps. The stagnating figures in the automotive industry, on the other hand, show that the number of processes to be automated by classic industrial robotics has been exhausted. For the automation of logistics processes, which are characterized by flexibility and dynamic changes with a high variance, autonomous and intelligent systems with a high level of adaptability instead of standard automation technology are therefore necessary.

Initial approaches within the framework of these developments showed that such autonomy could be achieved by using artificially intelligent algorithms. Despite the current euphoria in this area, however, there are also critical limitations. Thus, error susceptibility, as well as algorithms that exhibit unpredictable behavior, can cause severe damage in the wrong applications such as autonomous vehicles. To avoid such happenings, robust frameworks are necessary, like the one described in this article.

To provide a basic understanding of the processes and their relevant industrial influences, the following section focuses on logistics and the occurring objects therein.

2 Logistics

The overall goal of logistics is the fulfillment of the so-called 6Rs: the provision of

- the right amount
- of the right objects
- to the right destination
- at the right time
- in the right quality
- and for the right price.

For this purpose, all internal and cross-company flows of goods and information are planned, controlled, coordinated, carried out, and monitored.

2.1 Intralogistics

Intralogistics comprises the organization, control, implementation and optimization of internal material flows, information flows, and goods handling in industry, trade, and public institutions.

To achieve the goals mentioned in the previous section, the conveying and storage technology and complete systems used must be adapted to the product properties, such as size, shape, weight, and sensitivity of the material to be handled. This is achieved in logistics planning. Also of importance is the creation of a high degree of flexibility to adapt quickly to changing frameworks and environmental conditions. Nevertheless, costs must be continuously reduced, and processes must be carried out with a high degree of reliability. Process errors in intralogistics can lead to a shutdown of the assembly line and thus to high costs.

In addition to the production-synchronous provisioning processes, which have already been increasing over the past years in particular for large assemblies such

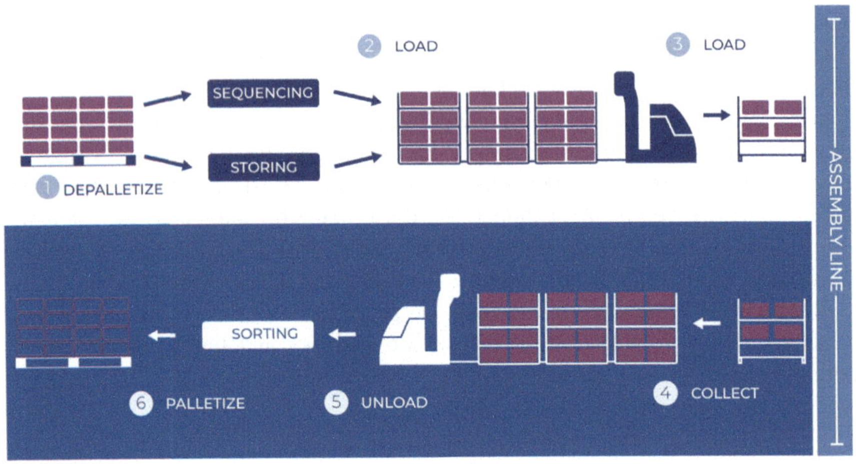

Fig. 1 Necessary handling steps in the intralogistics process chain (own illustration)

as seats or the instrument panel, the majority of the required components are still delivered to the assembly lines in asynchronous production. This material flow—which is shown in Fig. 1—is the focus of this work and will, therefore, be examined in more detail below.

The components to be assembled are delivered in containers on pallets by the suppliers on trucks. After the trucks have been unloaded, the pallets are transported to the goods receiving area. There the individual containers are depalletized (1) and stored in an automated small parts warehouse. As soon as the components contained in the container are needed, they are removed from storage and delivered to the assembly line (3). Since the space available on the assembly line is limited, not all containers can be provided there. Therefore, some components have to go through the sequencing steps. The components there are getting pre-sorted for the next vehicles to be assembled and then transported separately to the provision areas. Transport—direct or indirect—to the assembly line is currently mainly carried out by tugger trains. If the containers are directly from the small parts warehouse, the shelves on the tugger train trailers are filled automatically. Arrived at the assembly line, the containers are removed by the tugger train driver and transferred to the staging racks. In return, the empty containers are collected and reloaded onto the tugger train (4). The empties collected in this way are then sorted on conveyor systems in the empties sorter and palletized by type (6). Finally, these pallets are prepared for transport and sent back to the suppliers. Since the explanations in this doctoral thesis concentrate on the handling of containers, the following handling steps are the focus of the following sections:

- Depalletizing of full containers;
- Removal and provision of containers from one shelf to another;
- Palletizing of empties.

These process steps are analyzed in more detail below. The following explanations are based on process analyses in several car factories.

2.2 Objects

This section describes the objects required for material flow in more detail. Those are Containers, Pallets as well as shelves.

2.2.1 Containers

As already mentioned, the components to be assembled are delivered to the production facilities by the suppliers as complete loading units. A loading unit is the combination of several load carriers that are referred to as containers in the following, with one loading aid. These can be differentiated by size (small load carriers (container) and large load carriers) and degree of specialization (standard load carriers special load carriers). The main functions of load carriers and the formation of loading units are the protection of materials and the optimization of transport, handling, and storage processes in logistics. The BMW Group currently uses more than 3,000 different containers due to the specific requirements of the components and the material flow. A large number of different individual containers reflect the trend toward increasing model diversity in the automotive industry. Special containers are specially developed and constructed for the transport and storage of components with complex part geometry or specific requirements. BMW uses approx. two hundred fifty different special containers per vehicle derivative.

By standardizing container properties, standard containers can be used in all plants and technologies and throughout the supply chain. The specified dimensions are often based on DIN or VDA standards. Unlike special containers, standard containers are not assigned to any specific components and can, therefore, be used flexibly.

The main requirement for the containers from logistics is the damage-free delivery of faultless parts to the installation site. Not only the protective function of the containers plays a decisive role, but also the quality of the load carriers. Containers must absorb static and dynamic forces along the logistics chain. In almost all handling processes, this is reflected in the quality of the containers. Furthermore, the components to be transported have an immense influence on the quality of the container. If oily parts have to be delivered in a container, residues remain inside the container after removal at the production site. If a container passes through the container circuit several times, heavy soiling and wear can occur. These described phenomena lead to some properties which are of particular importance in connection with the automation of these processes. They are addressed in the following.

Optically, the high visual variance is particularly relevant. Depending on the vehicle derivatives to be produced, up to 400 different container types can be observed per plant. These differ mainly in terms of color, dimensions, and the material used.

Fig. 2 Visualization of the planned container variety (own illustration)

Fig. 3 Different optical appearances of only one container type (unplanned variety) (own illustration)

Figure 2 shows a small part of this planned variance. Due to the harsh industrial conditions prevailing in logistics, such as dust or other contamination, the effect of high physical forces due to frequent handling steps and the continuous loading and unloading of the containers so that humans can identify them, an unplanned optical variance can be observed in the plants in addition to the planned variation. This is shown in Fig. 3.

Another characteristic feature of containers is the different distribution of containers in the material flow caused by standardization efforts. This is shown in Fig. 4. This illustration also shows the plant dependency of the containers used. This is actively induced by the derivatives to be produced. Geographical aspects also play a role if, for example, in overseas plants, a large proportion of the materials have to be delivered by sea freight.

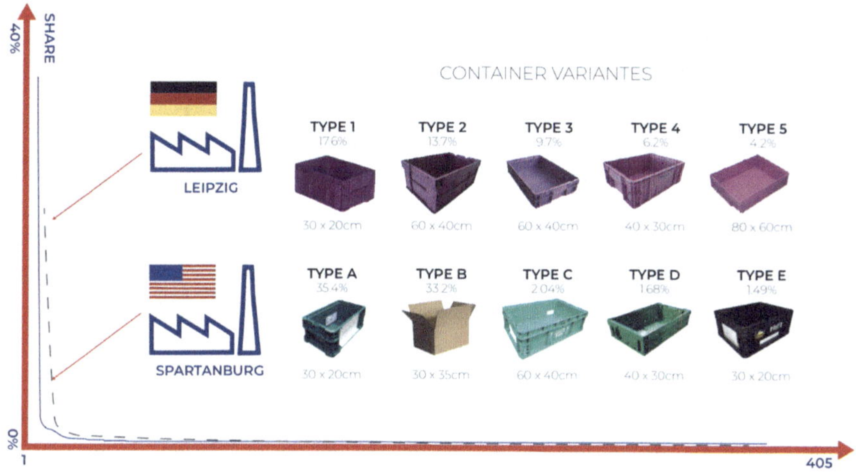

Fig. 4 Container distribution in the entire materialflow in different plants (own illustration)

The robots that should be used to automate those handlings steps have to deal with this whole variety. Therefore intelligent vision systems are necessary. The resulting algorithm is focused on the following section.

3 Perception Algorithm

In this section, the Perception Algorithm is presented. Following the description of its basic concept, the three submodules are explained in further detail.

3.1 Basic Concept

Research in the field of perception of robots in real environments is still relatively young, and so far, no "silver bullet" has been able to establish itself. This explains the variety of basic approaches to this topic, starting with the sensor input data up to their further processing in the different algorithms. This is illustrated in Fig. 5 and is described in more detail below. At this point, it should be added that many of the cited sources address only part of the overall perception processing process.

The starting point for this process is real or sometimes even a simulated scene with objects. Using appropriate sensors, the robot can capture these in the form of two- or three-dimensional images (RGB or RGB-D) and point clouds. These go through a multistage process before the final gripping point can be determined.

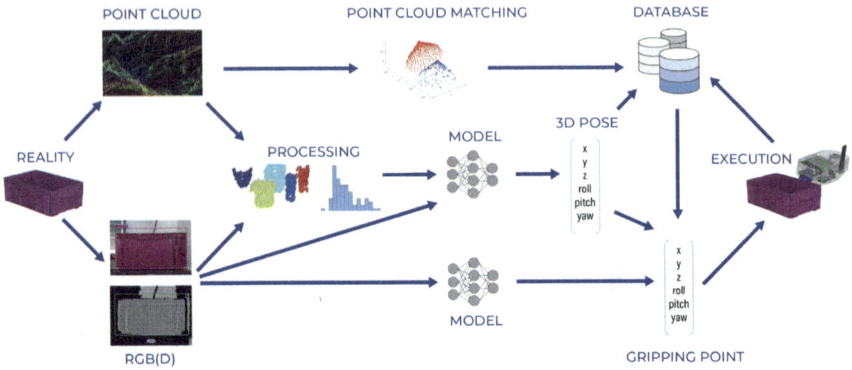

Fig. 5 Overview about different approaches for gripping pose detection (own illustration)

One significant intermediate result is the determination of the 3D pose of the object. The recorded input data is first segmented before the released objects can be classified. Following on from this, the precise position of the object in space can be estimated. Alternatively, some approaches predict the spatial position based on two-dimensional RGB images through neural networks without the mentioned explicit intermediate stages. Another method is point cloud matching, which attempts to place three-dimensional models of various objects in the captured point clouds.

If the 3D pose of an object is known, the gripping point can be determined. A detailed examination of the state of research reveals two approaches: analytical and empirical. In the latter approach, the gripping point is determined by comparing the 3D pose of the object with already stored gripping scenarios from databases. For an analytical determination, the areas available on the segmented object are examined about various criteria necessary for successful grasping. The criteria depend on various influencing factors, such as the surfaces of the objects to be gripped or the type of gripper (finger or suction gripper).

Once this gripping point has been determined, the manipulation task can be executed by the robot. The complete gripping process can then be evaluated concerning its success. This generated knowledge can be used either directly for retraining the model or as a further example in a database.

In addition to the subdivided approaches described above, there are further attempts to predict the most suitable gripping point via neural networks or a combination of these. In particular, procedures via reinforcement learning and reinforcement learning combined with imitation learning have received increased attention in recent years.

To overcome the complexity and error susceptibility of existing holistic algorithms, a modular approach is chosen in the described application case. Individual modules are sought for the core functions for processing the input data up to the output of the gripping point. These have the advantage that, on the one hand, there is a much greater choice of solution and, on the other hand, these have existed for some time, which leads to more reliable statements regarding performance and efficiency.

For example, it can also be seen from the upper picture that there is considerably more work available to determine the gripping coordinates from the 3D pose of the object than with the holistic approaches described. According to the described application case, the perception algorithm for the autonomization of the depalletizing in the incoming goods department as well as the one for the provisioning robot must fulfill the following partial tasks:

- Identification of the searched objects;
- Selection of the next required object in the process execution;
- Determination of the gripping point of this object in space.

Accordingly, the perception algorithm can be put together as follows (Fig. 5). In the following subsections, the selection or adaptation to the use case will be discussed in more detail, belonging to the individual submodules.

3.2 Module 1: Detection

This module aims to identify and locate objects of the searched classes in the field of vision of the robot. Congruent to the general requirements of the Perception Algorithm, the following aspects must be fulfilled. The detection module has to

- have the necessary precision to provide good input data for the following modules,
- be robust enough to enable stable process execution independent of dynamically changing environmental influences,
- be fast enough in execution so that the robot can meet the cycle time requirements for the respective handling steps, and
- be intuitive and simple enough to be adapted and implemented with manageable manual effort.

For this, a module is required, which outputs the position of the searched objects based on two-dimensional color images.

No best practice path has yet emerged for the object recognition frameworks. Instead, there are several general approaches in the literature which achieves comparable results but determine them in different ways. Simplifying these can be divided into two groups: network combinations and single-shot detectors. Exemplary, only two possible solutions will be discussed here on a superficial basis. These are Faster R-CNN's representative for network combinations and SSD, which represent the cluster group of the same name. For the former, potentially relevant image regions are first identified. These are then analyzed in a network. A characteristic feature of this network architecture is that in addition to the classification output, a Fully Connected Layer with softmax is implemented as a parallel Fully Connected Layer, which represents the Bounding Box Regressor. In contrast, the so-called Single-Shot Detectors do not require any image region suggestions. A single network is used,

Fig. 6 Target-specification for the Detection Module (own illustration)

which uses different bounding boxes, which are then adapted as part of the prediction. This procedure leads to significantly higher speed with comparable performance (Fig. 6).

To create a bigger decision basis, three architectures are chosen for comparison. These three are the following:

- YOLO,
- SSD, and
- Faster R-CNN.

YOLO was chosen because of its high frame rate and the consideration of the overall context of the image in object recognition, faster R-CNN because of its high precision and robustness, and SSD because it has the same one-shot learning advantages as YOLO, but in theory performs better on close objects.

One essential key for the training of a deep learning-based object detector is the dataset creation. This step is very time consuming, since preparing the dataset for the training requires acquiring a large number of representative samples and hand-labeling of the depicted objects and their positions in the image. The labeled images represent the ground truth, which should correspond with the range of scenarios in which the object detector should operate. The ground truth should contain images of different objects from the application's environment of the object detector from different poses and conditions. In this use case, the images are collected according to the described logistics areas.

To gather enough samples for the training, around 4.000 images from different plants of BMW Group in Germany (Munich, Leipzig, and Regensburg) were collected. Moreover, seven object classes were labeled, namely container, assembly line shelf, tugger train shelf, dolly, wheel, cage box, and pallet.

Faster RCNN

The pretrained *Faster R-CNN Resnet50* architecture provided by the *TensorFlow API* is used. The data is converted to the standard *TensorFlow* format as TFrecords. The best performance of the network is recorded with the following fine-tuned training configuration and hyperparameters, presented in the Table 1 (Fig. 7).

Table 1 Training hyperparameters for Faster RCNN

Parameter	Value
Image resizer	Yes
Data augmentation	Yes
Batch size	16
Momentum	0.9
Initial learning rate	0.0003
Learning rate	Step decay

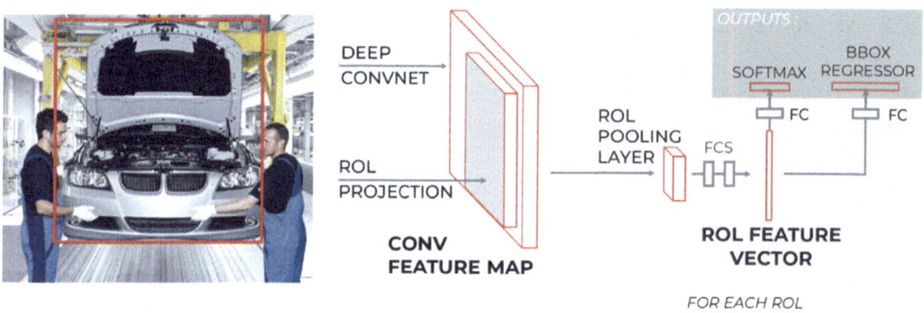

Fig. 7 Function principle of Fast R-CNN

Table 2 Training hyperparameters for SSD

Parameter	Value
Image resizer	Yes
Data augmentation	Yes
Batch size	12
Optimizer	RMSprop
Momentum	0.9
Initial learning rate	0.004
Learning rate	Exponential decay

SSD

Different models based on SSD architecture were tested. The *SSDLite MobileNetV2* architecture available on the *TensorFlow API* platform achieved the best performance. For the training, the TFrecords format for the data is required. The configuration and hyperparameters used for the training are shown in the Table 2 (Fig. 8).

YOLO

The last version of YOLO, *YOLO V3* implemented with the *Darknet* framework is fine-tuned and trained. The Table 3 shows the configuration and hyperparameters used during the training.

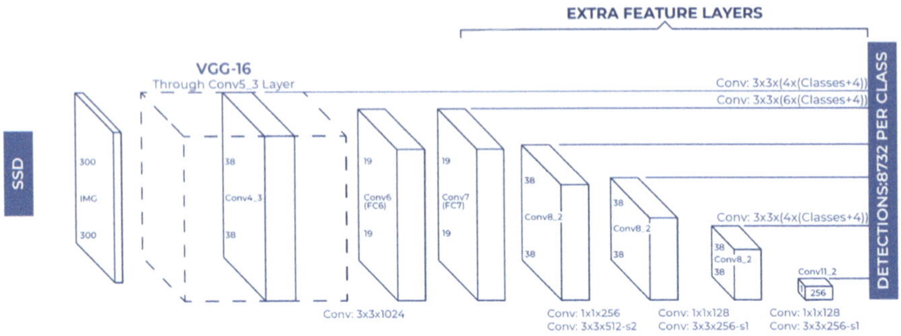

Fig. 8 Architecture of SSD

Table 3 Training hyperparameters for YOLO

Parameter	Value
Image resizer	Yes
Data augmentation	Yes
Batch size	12
Momentum	0.9
Initial learning rate	0.001
Learning rate	Step decay

As a guideline, the training process is considered completed when the loss reaches a constant value and stops decreasing. The time needed for the training of *Faster R-CNN resnet50*, *SSDLite MobileNetV2 coco* and *YOLO V3* is, respectively, 12h, 9h, and 48h. The focus of the robotics applications described here is primarily on the highest possible reliability and accuracy of results in order to guarantee the robustness of the processes in the industrial environment. Therefore, the training time itself was not further optimized during these training sessions. Particularly with the YOLO training sessions, significantly shorter training times could have been achieved through larger learning rates—but with lower overall efficiency (Fig. 9).

Selection

To evaluate the performance of the object detectors, the test dataset is used. For each image in the test dataset, the predicted bounding boxes annotations and the time needed to process the image are stored. Each predicted bounding box is represented with five descriptors:

- The class of the object;
- The top left x coordinate;
- The top left y coordinate;
- The right bottom x coordinate;
- The right bottom y coordinate.

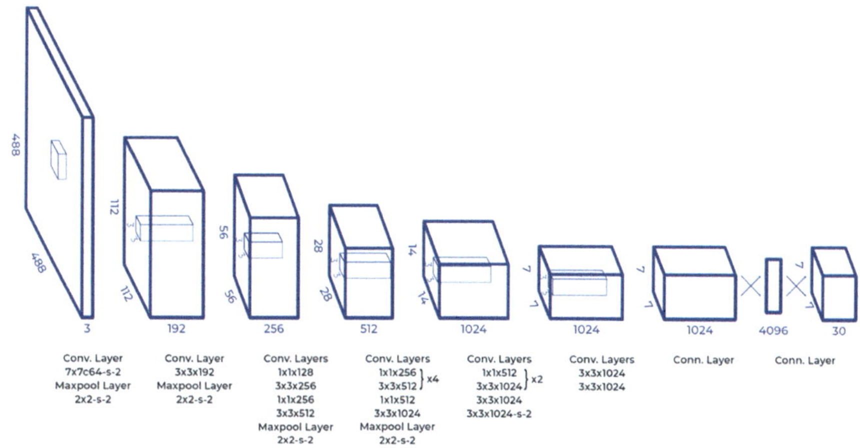

Fig. 9 Architecture of YOLO

Table 4 Performance of the object detectors over three different classes. The performance is measured by Average Precision per Class in %

Framework	Container	Pallet	Shelf
Faster R-CNN	87.1	68.0	82.4
SSD	70.8	59.2	69.0
YOLO	86.2	71.1	79.0

To measure the accuracy of the object detectors, the predictions on the test dataset are compared to the ground truth. Therefore, the mean average precision mAP with an IoU threshold equal to 0.5 is computed. The AP separately for each class and the overall mAP over all classes (7 classes) are both calculated. The first result is summarized in Table 4. Besides, the number of frames per second FPS is computed to evaluate the speed of the detectors.

The models achieve different performances over different classes. This can be explained by an imbalanced class distribution in the dataset and the fact that object detectors are sensitive to object's sizes and specific visual appearance. For instance, some object classes like container have more samples than the other object classes such as pallet. Therefore, for the comparison, the overall mAP is considered. In industrial applications, visual systems, used in robot motion control for material handling purposes, have to be precise and fast. To this end, the question is which detector and what configurations give us the best balance of speed and accuracy. Below is the overall accuracy versus speed trade-off of *Faster R-CNN Resnet50*, *SSDLite MobileNetv2 COCO*, and *YOLO V3*.

Inference results indicate that, in terms of precision and speed, YOLO is leading with the score of mAP equal to 76% and FPS value equal to 19.4. Faster R-CNN, on the other hand, has a good performance in terms of precision with mAP equal to 72% but was prolonged. Whereas, SSD was faster than Faster R-CNN but has a low precision with only 61% mAP. Based on its outperforming result in terms of precision and speed, YOLOv3 is suitable to be the object detector in the vision system and used for the Detection Module in the following.

To give further insights, the precision–recall curves of the respective classes for the selected network are shown in Fig. 10.

3.3 Module 2: Selection

The output of the first module of the perception algorithm is the set of objects detected in the corresponding image of the searched class. This is checked in the second module Selection concerning its process relevance so that finally an object remains which is to be gripped by the robot.

The scope of the module to be implemented will first be defined. These functions are then designed and implemented on the basis of an analysis of the current state of the art.

To do this, all objects that are not in the robot's workspace must first be removed from the result lists at this point. Next, the relevant object for the current gripping process must be selected from the remaining objects based on a processing strategy to be designed. It should be noted that it has to be ensured that the handling process neither damages other objects nor restricts the robot in the execution of the overall process. This could be the case, for example, when depalletizing, if the robot grabs the containers in such an order that it can no longer pull the following containers over the previous ones (Fig. 11).

The functional explanation of this module described here is very specific to the topic to be solved in the context of the present work. Due to this specialization, no overall algorithms or algorithm modules could be found in the literature on which to build at this point.

Following the described objective, the module itself is divided into two core components, namely the

- Selection of the attainable physical objects for the robot (working space selection);
- Selection of the objects tangible for the current process step (process step selection).

3.3.1 Workingspace Selection

The number of physically accessible objects for the robot is limited on the one hand by the robot hardware used and on the other hand, by the infrastructure involved. During depalletizing, the detected objects are limited by the pallet height and the maximum layer height in the y-direction. Since all objects to be depalletized must be on the palette, their length represents the x-direction boundary.

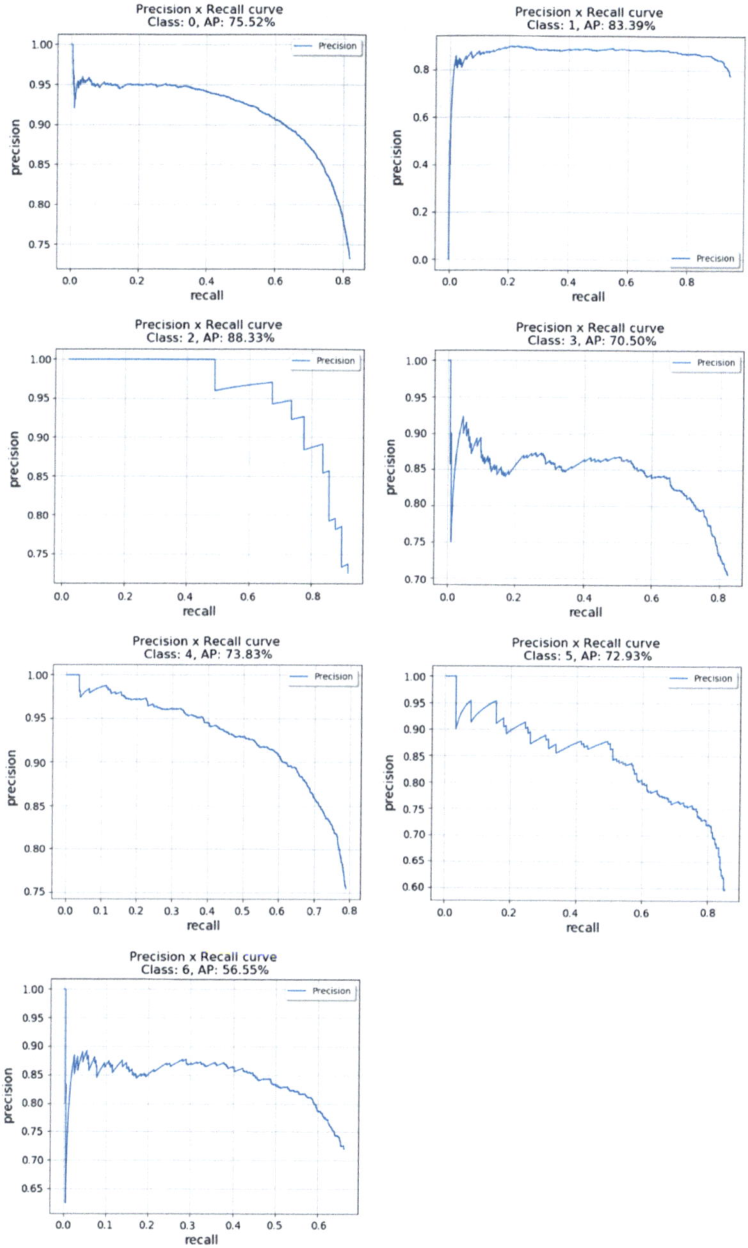

Fig. 10 precision–recall curves of the respective classes: container, shelf, tugger train shelf, dolly, wheel, cage box, palette (own illustration)

Fig. 11 Target-specification for the Selection-Module (own illustration)

Fig. 12 Example for applied workingspace selection (own illustration))

In the case of belt provision, the objects to be handled must fulfill the general condition that they are within the tugger train trailer rack when full loads are provided. In return, when empty containers are collected, the empty containers must be part of the relevant staging rack. Besides, there is the working instruction to better differentiate between full and empty containers. The latter is only provided in the lowest shelf line for the tugger train collection (Fig. 12).

During the final palletizing of the empty containers, only containers located on the conveyor system may be used for determining the gripping point. Here, the respective delivery stitch must be delimited so that two potentially alongside robots do not block each other.

3.3.2 Process Selection

As with workspace selection, the criteria for process selection must be postulated robot-selectively. In general terms, decisions must be made in two categories, namely based on the sequence of grips and the behavior in the event of an error. Whereas the former involves deciding the sequence to be processed, the latter specifies how to deal with mistakes.

In contrast to the supply of the assembly line or palletizing of empties, process selection during depalletizing at goods receipt is of particular importance. It must be ensured under all circumstances that the robot depalletizes in such a way that it continues to be able to retract the other containers after each container has been

Fig. 13 Example for applied process selection (own illustration))

removed. The layer pattern to be processed for this is as follows. At first, the detected object is started in the upper left corner, and the first row is removed, as shown in Fig. 13. Then the containers behind them—again from left to right—are drawn in one after the other. Once the entire first layer has been removed, you can start with the layers below. These are processed according to the same scheme until the pallet is entirely depalletized.

Since this sequence must be adhered to during depalletizing, no special measures such as gripping alternative containers in the event of errors can be actively carried out by the robot. Therefore, an employee must be called to solve the problem so that the robot can continue to work according to its predefined scheme.

The process does not have to adhere to a defined sequence when providing the assembly line. Although the containers are drawn into the robot via the linear axis system as in depalletizing, the containers do not have to be moved one above the other for this as they are located in roller-driven shelves. In this case, the decision as to which container is to be gripped can be made based on two decision criteria. On the one hand, it is possible to grab the container next, where the perception module has the highest confidence. On the other hand, it is also possible to process the containers in such a way that the robot's paths and thus the unloading time of the tugger train can be minimized. For this, a further vision module is necessary to identify the actual container contents. This can be achieved by reading the barcodes on the container. By the corresponding fixed allocation to the rack storage bins, the constant tuggers and the optimal unloading strategy can be determined.

In the case of false detections or even incorrect grips, it is also possible to continue working with the next most relevant container, since the higher level process execution is not negatively affected in the long term.

When palletizing objects, the decision is made primarily concerning the confidence of the object recognition. Since the objects are rearranged in the working area of the robot by the roller conveyor after each container handling operation, no restrictions about the processing sequence are necessary here. One phenomenon that can be observed more frequently here, however, is intangible containers. This can be caused, for example, by residual packaging from the components transported in the containers. Also, partial soiling may occur, or labels may still be present in the containers. If this happens, the specific gripping point is no longer taken into account in the selection for the next iteration, even if the confidence contradicts this.

Fig. 14 Target-specification for the Localization-Module (own illustration)

3.4 *Module 3: Localization*

After the first two modules of the Perception Algorithm have detected the searched objects and selected the object relevant for further process execution, the third step is to determine the exact gripping pose. First, the exact target of the module is specified in more detail. Subsequently, relevant approaches are selected based on an analysis of existing methods and theoretical considerations. These are then adapted and implemented to the framework conditions required here before the final selection is made.

The target of the Localization Module is to determine the gripping pose in three-dimensional space as it is visualized in Fig. 14. Depending on the application, other coordinate vectors may be required. For example, when depalletizing, the object rotations can be set to zero, since it can be assumed that the containers are positioned horizontally and vertically on the pallet, as otherwise, it would not be possible to transport the objects at all. Besides, small tolerances are no problem as they can be compensated by the flexible plastic suction cup of the gripper. When providing the objects, the gripper must be adapted to the new position of the shelves. Likewise, rotations of the container itself can occur in the roller-driven gravity racks. These must be able to be recognized. When palletizing empty containers, in addition to the x, y, and z positions, the rotation of the container must also be taken into consideration.

It is crucial for the determination of the gripping point that neither significant contamination nor roughness occurs on the selected surface regions, as this would impede the sealing of the suction pad. With suction pads, the force that can be applied is directly proportional to the size of the suction pad. To enable a stable gripping process, this was, therefore, dimensioned as large as possible. However, this means that a very high degree of precision is required to determine the gripping pose, as otherwise, the gripper would rest on the webs on the surface of the container. Since, also, it is only possible to grip on the container surface or, during palletizing, even on the inside of the container, it is necessary to ensure that the identified gripping surface does not contain a label or other sticker.

In order to meet the requirements described for the respective gripping point and thus to be able to rule out damage due to incorrect gripping, the most suitable

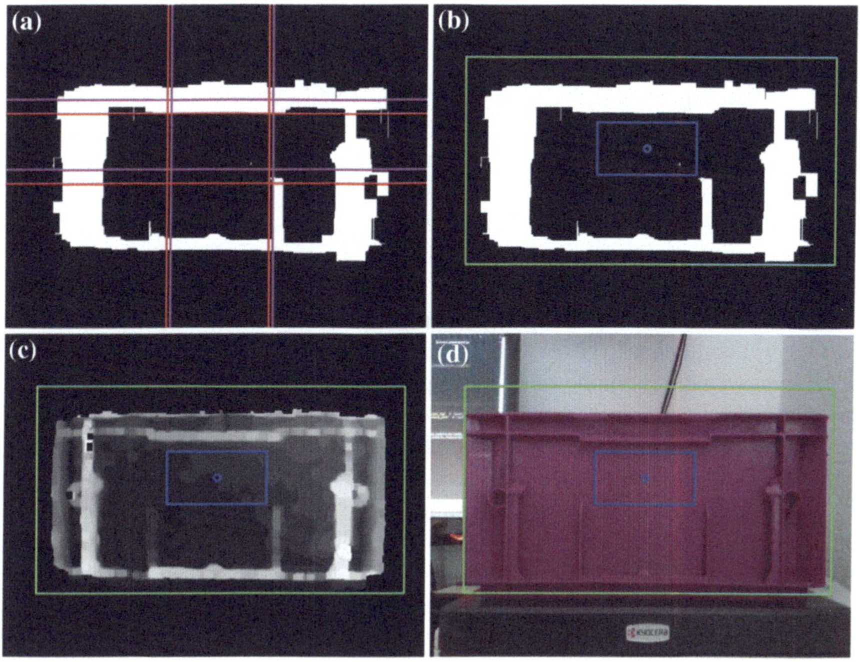

Fig. 15 Final execution of gripping area analysis (own illustration). **a** Searching for a rectangle, **b** Suction cup area, **c** Suction cup area, drawn in the original depth image, **d** Suction cup area, depicted in original RGB image

gripping point is determined in this module on the basis of depth data. For this purpose, the depth image passes through several filters. With these, erroneous data can be added and any obstacles that prevent robust gripping can be avoided. Following these preparatory image processing steps, suitable gripping surfaces can be searched for in the free areas on the top side of the container. These final steps are shown in Fig. 15.

This approach was chosen because its generic nature makes it possible to determine safe gripping points regardless of container type and material. Further approaches were also tested. For example, the approach of determining the gripping point by predefined offsets from the bounding box boundaries of the objects detected in the first module was not used. However, this increases the dependency of the entire algorithm on the neural networks and thus the uncertainties in the overall system. A further approach was to add the gripping surfaces as separate classes for training in Module 1. Due to the large variety of containers and the numerous special cases (label on the container, damage, …), no promising results could be achieved with the number of data used (almost 3500).

4 Conclusion

The algorithm explained in this section is currently being evaluated in a depalletizing robot.

The following results were achieved during tests in the real process environment at the Leipzig vehicle plant (Fig. 16).

It should be noted that thanks to the perception algorithm developed in the context of this publication, gripping processes could be carried out under significantly more difficult industrial conditions with a success rate of 72%. The requirements derived from the use case at the beginning could thus be achieved in comparison to state of the art. Only reliability has to be increased. This is realistic due to the modular approach, as the improvement is composed of the improvements of the individual modules. On the other hand, the improvement of a neural network of this magnitude would hardly be conceivable. It should also be noted at this point that the 28% of unsuccessful grips are also due to the interaction between the newly developed robot hardware and the freshly developed perception algorithm. Due to the small tolerances, phenomena can occur such that, for example, the gripping point fits perfectly, but the accuracy of the robot system collides with the edge of a box due to vibrations or inaccuracies. Noisy sensor data also play a role here. At the end of the experiments, the robot's reaction to faulty grips, which can be classified by evaluating the vacuum on the gripper, was specified as that it should try again. In the second attempt, just under 50% of the objects that had not been touched before could be gripped stably. This means that the robot initially satisfies the framework conditions and can continue to be operated on-site in the real series process.

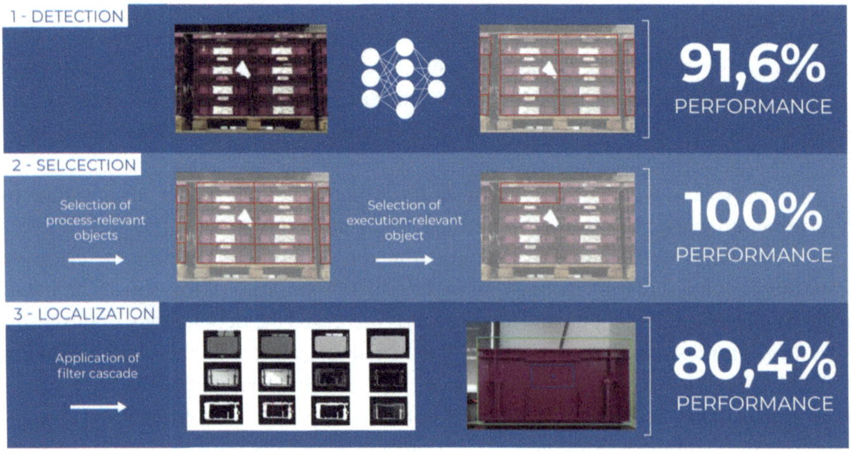

Fig. 16 Summary of the evaluation of the Perception Algorithms functional modules (own illustration)

In addition, these results show the strong dependence of the training data with the overall results to be achieved afterward in the series process. As a glance at the precision–recall curves also shows, very good results can be achieved, for example, with the containers class, which account for a high overall proportion of the material flow. Rarer containers as well as objects that are less visible due to their function (a pallet for up to 100 containers) perform worse. This fact is particularly important for future steps, for example, when the robot continuously creates and learns from training data. Other imponderables such as the change in the overall performance depending on the environment (other factory halls, other plants) must also be investigated in further steps. The operation of the robots in series production generates a great deal of additional data, which provides a solid basis for further improvements to the deep learning algorithms and thus enables the sustainable use of additional robot solutions in plants worldwide.

References

1. K. Permenter, Cost and Complexity: Top Challenges of Today's Chief Supply Chain Officer (Aberdeen Group, 2012)
2. A. Stringflow, How to Reduce Logistics Costs: 19 Experts Reveal Ways Organizations Can Cut Their Logistics Transportation and Carry Costs (New Technology, 2018)
3. D. Arnold et al., Handbuch Logistik (Springer, Berlin, 2008), ISBN: 978-3-540-72929-7
4. R. Hodson, How robots are grasping the art of gripping. Nature **557**, 23–25 (2018)
5. D. Holz, Registration with the point cloud library. IEEE Robot. Autom. Mag. **22**(4), 110–124 (2015)
6. A. Romea et al., Object recognition and full pose registration from a single image for robotic manipulation, in *ICRA* (2009)
7. M. Dogar et al., Physics-based grasp planning through clutter, in *RSS* (2012)
8. I. Lenz et al., Deep learning for detecting robotic grasps, in *ICLR* (2013)
9. A. ten Pas et al., Grasp pose detection in point clouds (2017). arXiv: 1706.09911v1
10. W. Kehl et al. SSD-6D: making RGB-based 3D detection and 6D pose estimation great again, in *Proceedings of the IEEE International Conference on Computer Vision* (2017)
11. F. Spenrath et al., Gripping point determination for bin picking using heuristic search. Procedia CIRP **62**, 606–611 (2017). https://doi.org/10.1016/j.procir.2016.06.015
12. R. Bostelman et al., Survey of industrial manipulation technologies for autonomous assembly applications. NIST-Internal Report 7844 (2012)
13. M. Robertson et al., New soft robots really suck: vacuum-powered systems empower diverse capabilities. Sci. Robot. **2**(9) (2017)
14. S. Gu et al., Deep reinforcement learning for robotic manipulation with asynchronous off-policy updates, in *2017 IEEE International Conference on Robotics and Automation (ICRA)* (2017). IEEE
15. L. Fanet al., SURREAL: open-source reinforcement learning framework and robot manipulation benchmark, in *Conference on Robot Learning* (2018)
16. J. Mahler et al., Dex-Net 3.0: computing robust robot vacuum suction grasp targets in point clouds using a new analytic model and deep learning (2017). arXiv:1709.06670
17. A. Zeng et al., Multi-view self-supervised deep learning for 6d pose estimation in the amazon picking challenge, in *Robotics and Automation (ICRA)* (2017)
18. S. Kai-Tai et al., CAD-based pose estimation design for random bin picking using a RGB-D camera. J. Intell. Robot. Syst. **87**(3–4), 455–470 (2017)

19. J. Tremblay et al., Falling things: a synthetic dataset for 3D object detection and pose estimation (2018). arXiv:1804.06534
20. H. Hiteshree et al., Analysis of feature based object mining and tagging algorithm considering different levels of occlusion, in *Communication and Signal Processing (ICCSP)* (2017)
21. S. Sundararajan et al., Continuous set of image processing methodology for efficient image retrieval using BOW SHIFT and SURF features for emerging image processing applications, in *2017 International Conference on Technological Advancements in Power and Energy (TAP Energy)* (2017)
22. W. Li et al., Visual recognition in RGB images and videos by learning from RGB-D data. IEEE Trans. Pattern Anal. Mach. Intell. **40**(8), 2030–2036 (2018)
23. R. Shaoqing et al., Faster R-CNN: towards real-time object detection with region proposal networks. IEEE Trans. Pattern Anal. Mach. Intell. **6**, 1137–1149 (2018)
24. R. Shaoqing et al., Object detection networks on convolutional feature maps. IEEE Trans. Pattern Anal. Mach. Intell. **39**(7), 1476–1481 (2017)
25. J. Redmon, A. Farhadi, Yolov3: an incremental improvement. arXiv:1804.02767
26. Liu, Wei, et al. Ssd: Single shot multibox detector, in *European Conference on Computer Vision* (Springer, Cham, 2016), pp. 21–37
27. R. Girshick, Fast r-cnn, in *Proceedings of the IEEE International Conference on Computer Vision* (2015), pp. 1440–1448

Author Index

A
Abodo, Franklin, 75
Arif Wani, M., 1

B
Berthaume, Andrew, 75
Blesing, Christian, 137

C
Chakraborty, Shayok, 113
Cheung, Albert C., 53
Concha, Darwin Ttito, 95

D
de Almeida Maia, Helena, 95
de Lima Chaves, Hugo, 95
de Souza Brito, André, 95

F
Friedrich, Christoph M., 137

G
Gan, Min, 53
Goehring, Daniel, 155

I
Iglovikov, Vladimir I., 39
Irrenhauser, Thomas, 155

K
Kalinin, Alexandr A., 39
Kantardzic, Mehmed, 1

M
Marica, Roummel F., 9

P
Panchanathan, Sethuraman, 113
Pedrini, Helio, 95
Poss, Christian, 155
Prueglmeier, Marco, 155

R
Rafati, Jacob, 9
Rakhlin, Alexander, 39
Ranganathan, Hiranmayi, 113
Rittmuller, Robert, 75

S
Salehi, Vahid, 155
Sayed-Mouchaweh, Moamar, 1
Schweigert, Anneliese, 137
Shen, Yanyan, 53
Shvets, Alexey A., 39
Sumner, Brian, 75

T
Tacon, Hemerson, 95

M. A. Wani et al. (eds.), *Deep Learning Applications*,
Advances in Intelligent Systems and Computing 1098,
https://doi.org/10.1007/978-981-15-1816-4

V

Venkateswara, Hemanth, 113

Vieira, Marcelo Bernardes, 95

Villela, Saulo Moraes, 95

W

Wang, Hongfei, 53
Wang, Shuqiang, 53

Z

Zoghlami, Firas, 155